Asia International Conference on Energy and Environment Engineering (ACEEE 2018)

IOP Conference Series: Earth and Environmental Science
Volume 133

Shanghai, China
19 – 21 January 2018

ISBN: 978-1-5108-6151-0
ISSN: 1755-1307

Printed from e-media with permission by:

Curran Associates, Inc.
57 Morehouse Lane
Red Hook, NY 12571

Some format issues inherent in the e-media version may also appear in this print version.

Printed by Curran Associates, Inc. (2018)

For permission requests, please contact the Institute of Physics
at the address below.

Institute of Physics
Dirac House, Temple Back
Bristol BS1 6BE UK

Phone: 44 1 17 929 7481
Fax: 44 1 17 920 0979

techtracking@iop.org

Additional copies of this publication are available from:

Curran Associates, Inc.
57 Morehouse Lane
Red Hook, NY 12571 USA
Phone: 845-758-0400
Fax: 845-758-2633
Email: curran@proceedings.com
Web: www.proceedings.com

Table of contents

Volume 133

2018 Asia Conference on Energy and Environment Engineering (ACEEE 18)

19–21 January 2018, Shanghai, China

Accepted papers received: 9 March 2018
Published online: 10 April 2018

Ecological Construction and Environmental Assessment

IOP Conf. Series: Earth and Environmental Science **133** (2018) 011001 doi:10.1088/1755-1315/133/1/011001

Preface

It is our pleasure to introduce you the Proceedings of 2018 Asia Conference on Energy and Environment Engineering (ACEEE 2018) which will provide a highly competitive forum for reporting the latest developments in the research and application of Energy and Environment Engineering. ACEEE 2018 will offer an extensive program of interest to academia, government and industry. A series of great presentations will be presented to improve your skills and the relevant professional practices.

The objective of ACEEE 2018 will always be bringing together a large group of researchers, scientists, academics, and engineers in the area of Energy and Environment Engineering from all over the world to share ideas not only the technical interests, culture and history.

This proceedings of ACEEE 2018 presents papers submitted to the conference from universities, research institutes and industries. The papers selected depended on their quality and their relevancy to the conference. The volume tends to present to the readers the recent advances in the field of renewable energy, energy management and economic strategy, energy development and generation system, design and development of new energy vehicles, power and electrical engineering, system modeling and performance evaluation etc.

We would like to thank all the authors who have contributed to this volume and also to reviewers, speakers, chairpersons, sponsors and all the conference participants for their hard work and support to ACEEE 2018.

ACEEE 2018 Organizing Committee
January 30, 2018

IOP Conf. Series: Earth and Environmental Science **133** (2018) 011002 doi:10.1088/1755-1315/133/1/011002

Conference Committee

General Chairs
Tyrone Fernando, University of Western Australia, Australia

Zhixin Wang, Shanghai Jiaotong University, China

Technical Program Chairs
Hongqi Wang, Beijing Normal University, China

Hee-Je Kim, Pusan National University, Korea

S. M. Muyeen, Curtin University, Australia

Dylan Lu, University of Technology Sydney, Australia

Lin Chen, Tohoku University, Japan

Technical Committee Members
Wang Chengmin, Shanghai Jiao Tong University, China

Wang Jie, Shanghai Jiao Tong University, China

Feng Donghan, Shanghai Jiao Tong University, China

Marek Balazinski, École Polytechnique de Montréal, Canada

Bingtao Liu, North China University of Water Resources and Electric Power, China

Kamal Abdel Radi Ismail, University of Campinas, Brazil

Ock Taeck Lim, University of Ulsan, South Korea

Ab Halim Abu Bakar, University of Malaya, Malaysia

Nallusamy Nallusamy, SSN College of Engineering, India

Khamaruzaman Wan Yusof, Universiti Teknologi PETRONAS Malaysia

C.X Chen, Duangxi University, China

Gao Qiang, Shanghai Jiao Tong University, China

Fangyu Fan, Southwest Forestry University, China

Zhu Li, Shanghai Jiao Tong University, China

Luo Xiang, Shanghai Jiao Tong University, China

Jahangir Hossain, Macquarie University, Australia

Eng. VALENTINA E. BĂLAŞ, Aurel Vlaicu University of Arad, Romania

Ahmed Al-durra, Khalifa University of Science and Technology, UAE

Ma Liangdong, Dalian University of Technology, China

Hassan Ali, University of Newcastle, Australia

IOP Conf. Series: Earth and Environmental Science **133** (2018) 011002 doi:10.1088/1755-1315/133/1/011002

IOP Conf. Series: Earth and Environmental Science **133** (2018) 011003 doi:10.1088/1755-1315/133/1/011003

Peer review statement

All papers published in this volume of *IOP Conference Series: Earth and Environmental Science* have been peer reviewed through processes administered by the proceedings Editors. Reviews were conducted by expert referees to the professional and scientific standards expected of a proceedings journal published by IOP Publishing.

Prediction of line failure fault based on weighted fuzzy dynamic clustering and improved relational analysis

Xiaocheng Meng [1, 2], **Renfei Che** [1, 2], **Shi Gao** [1, 2, 3] **and Juntao He**[1, 2]

[1]Key Laboratory of Power System Intelligent Dispatch and Control of Ministry of Education, Jinan, 250061, CHN

[2]School of Electrical Engineering, Shandong University, Jinan, 250061, CHN

[3]State Grid Hebei Maintenance Branch, Shijiazhuang, 050070, CHN

Abstract. With the advent of large data age, power system research has entered a new stage. At present, the main application of large data in the power system is the early warning analysis of the power equipment, that is, by collecting the relevant historical fault data information, the system security is improved by predicting the early warning and failure rate of different kinds of equipment under certain relational factors. In this paper, a method of line failure rate warning is proposed. Firstly, fuzzy dynamic clustering is carried out based on the collected historical information. Considering the imbalance between the attributes, the coefficient of variation is given to the corresponding weights. And then use the weighted fuzzy clustering to deal with the data more effectively. Then, by analyzing the basic idea and basic properties of the relational analysis model theory, the gray relational model is improved by combining the slope and the Deng model. And the incremental composition and composition of the two sequences are also considered to the gray relational model to obtain the gray relational degree between the various samples. The failure rate is predicted according to the principle of weighting. Finally, the concrete process is expounded by an example, and the validity and superiority of the proposed method are verified.

1. Introduction

As the information industry's rapid development, the world has been fully into the era of big data. At present, the power transmission and power distribution are the key points in the application of large data in power system[1]-[2]. The main aspect is the equipment fault diagnosis, that is, grid equipment status monitoring and early warning analysis.

In the data processing and feature mining of big data, the commonly used machine learning algorithms involve fuzzy clustering and gray relational analysis, etc. In the case of fuzzy clustering, the fuzzy clustering method is used in the literature [3-4] to cluster the original data, however, these fuzzy

IOP Conf. Series: Earth and Environmental Science **133** (2018) 012001 doi:10.1088/1755-1315/133/1/012001

clustering algorithms do not take into account the differences between the various factors, believe that the importance of the various factors between the indicators are equal, and this is clearly inconsistent with the actual. In literature [5-6], a method is proposed to solve the difference of factors. The article adopts the weighted way to distinguish the importance of the data, and discusses the problem where the weight should be added in the algorithm. In literature [7-8], some common algorithms of use of weight are discussed, and the application of each algorithm is put forward, and it is pointed out that the weighted weight should accord with the rationality analysis. In the case of gray relational analysis, the literature [9-10] points out the specific application of gray relational analysis in fault identification, index evaluation and equipment maintenance, but it does not improve the gray relational model itself. Through the above related literature, and combined with the collection of relevant historical data and comprehensive analysis of electrical equipment early warning, the coefficient of variation is determined as the weight, and the classification of historical data is carried out by weighted fuzzy dynamic clustering; then, based on the basic idea and basic nature of gray relational analysis, the concept of Deng's slope is proposed, and the gray relational analysis is improved by using the incremental composition ratio and composition difference of two sequences, and the gray relational degree between the various sequences is obtained, and finally through the idea of the weight to predict the failure rate of the line.

2. Determination of the weight

In this paper, through the comparison of common weight assignment methods, combined with rationality analysis, the final introduction of coefficient of variation to determine the weight of each meteorological indicators.

The coefficient of variation uses the degree of variation of the evaluated object to carry out the assignment of the weight, and the dynamic assignment of the index can be realized. In the meteorological factors, the factors that have a high degree of influence on the line fault often have the characteristics of high degree of dispersion and high volatility, that is, larger variance. Therefore, this paper combines variance and coefficient of variation to determine the weight of each meteorological factors.

In this paper, a total of 10 meteorological sequences were collected to evaluate the meteorological indicators of a total of four, c_{ij} indicates the j -th meteorological index of the i -th meteorological sequence, $i=1,2,3,\cdots,m$, $j=1,2,3,\cdots,n$, $m=10$, $n=4$。

The specific calculation steps are as follows:

- Calculate the mean square error of the i -th meteorological index

$$\begin{cases} S_j = \sqrt{\sum_i^m (c_{ij} - \bar{c}_j)^2 / m} \\ \bar{c}_j = \dfrac{\sum_{i=1}^m c_{ij}}{m} \end{cases} \tag{1}$$

Where $\bar{c}_j$ is the j -th meteorological index and S_j is the mean square error of the j -th meteorological index.

- Calculate the coefficient of variation for the j -th meteorological index.

$$\delta_j = \frac{S_j}{\bar{c}_j} \tag{2}$$

Where δ_j is the coefficient of variation of the j -th meteorological index.

- Normalize the coefficient of variation, get the weight of the first p meteorological indicators.

$$\begin{cases} A_j = \dfrac{\delta_j}{\sum \delta_j} \\ \sum_{j=1}^{n} A_j = 1 \end{cases} \tag{3}$$

Where A_i represents the weight of the j -th meteorological index.

3. Weighted fuzzy dynamic clustering

The traditional fuzzy dynamic clustering algorithm does not take into account the differences between the various factors, This paper uses a weighted approach to represent the different degrees of importance between the various factors.

The specific methods are as follows:

- Normalize the original meteorological data c_{ij} according to the deviation standardization method.

$$d_{ij} = \frac{c_{ij} - \min\limits_{1 \le j \le n}\{c_{ij}\}}{\max\limits_{1 \le j \le n}\{c_{ij}\} - \min\limits_{1 \le j \le n}\{c_{ij}\}} \tag{4}$$

- Calculate the weight of each meteorological factor according to the steps of the coefficient of variation

$$a = [a_1, a_2, a_3, \cdots, a_n] \tag{5}$$

- Do weighting after normaling data

$$f = \begin{bmatrix} d_{11} & d_{12} & \cdots & d_{1n} \\ d_{21} & d_{22} & \cdots & d_{2n} \\ \vdots & \vdots & \vdots & \vdots \\ d_{m1} & d_{m2} & \cdots & d_{mn} \end{bmatrix} * \begin{bmatrix} a_1 & 0 & 0 & 0 \\ 0 & a_2 & 0 & 0 \\ \vdots & \vdots & \vdots & 0 \\ 0 & 0 & 0 & a_m \end{bmatrix} \tag{6}$$

- Solve Fuzzy Similarity Matrix.

$$R_{ij} = \frac{\sum_{k=1}^{m} (f_{ik} \wedge f_{jk})}{\sum_{k=1}^{m} (f_{ik} \vee f_{jk})} \tag{7}$$

Where $\vee$ for the big operation, $\wedge$ for the small operation.

IOP Conf. Series: Earth and Environmental Science **133** (2018) 012001 doi:10.1088/1755-1315/133/1/012001

- Solve $t(R)$ according to square method

$$\begin{cases} t(R) = R^{2k} \\ R \to R^2 \to R^4 \to R^8 \to \cdots R^{2k} = R^{2(k+1)} \\ R^2 = R \circ R = (h_{ij})_{m\times n},\ h_{ij} = \bigvee_{k=1}^{m} (R_{ik} \wedge R_{kj}) \end{cases} \tag{8}$$

- Determine the optimal threshold λ.

In this paper, the F statistic is used to determine the optimal threshold.

The number of samples corresponding to λ is r, the number of samples of the j-th class is n_j, the average of the k-th variable in the j-th class is $\overline{x}_k^{(j)}$, and the average of the k-th variable of the sample is $\overline{x}_k$:

$$\begin{cases} \overline{x}_k^{(j)} = \dfrac{1}{n_j}\sum_{i=1}^{n_j} x_{ik}^{(j)},\ (k = 1,2,\cdots,m) \\ \overline{x}^{(j)} = (\overline{x}_1^{(j)}, \overline{x}_2^{(j)}, \cdots, \overline{x}_m^{(j)}) \end{cases} \tag{9}$$

$$\begin{cases} \overline{x}_k = \dfrac{1}{n}\sum_{i=1}^{n} x_{ik},\ (k = 1,2,\cdots,m) \\ \overline{x} = (\overline{x}_1, \overline{x}_2, \cdots, \overline{x}_m) \end{cases} \tag{10}$$

Define the F statistic as follows:

$$F = \frac{\sum_{j=1}^{r} n_j \left\| \overline{x}^{(j)} - \overline{x} \right\|^2 / (r-1)}{\sum_{j=1}^{r}\sum_{i=1}^{n_j} \left\| x_i^{(j)} - \overline{x}^{(j)} \right\|^2 / (n-r)} \tag{11}$$

F obey F distribution rule which the degree of freedom is $r-1$, $n-r$. For a given test level α, check the $F_\alpha(r-1, n-r)$ distribution table, get the critical value of F_α, if $F > F_\alpha$, then consider that there are significant differences between the various types. If the number of $F > F_\alpha$ is not just one, then compare the ratio of $(F - F_\alpha)/F_\alpha$ to choose the larger one.

4. Improvement of Gray Relational Analysis

Gray relational analysis is one of the important components of gray system theory, and it is a quantitative and comparative method for the development of gray dynamic process.

In the evaluation of the gray relational model, the four principles of the Deng's gray relation is the basic criterion of the test model, and it is also the basic idea to be followed in the gray relational analysis. But the defects of Deng's relational degree are obvious, including: the least absolute difference between the two levels and the maximum absolute difference between the two stages is serious; For a given resolution factor, the dimensionless processing is not warranted; it cannot meet the normative requirements.

In this paper, the above-mentioned nature is used as the basic idea to improve the gray relational analysis, and take into account the indicators among measuring the degree of similarity of the sequence curve, only slope is the indicator with inheriting order property, so combine Deng's relational method with the conception of slop, named Deng's slop ,to solve the problem of inheriting order property of Deng's slop. And also, this paper uses two series of incremental composition ratio and composition difference to define the relational coefficient, which can make full use of the information contained in the data sequence, which reflects the more real degree of relation between data sequence.

The specific definition is as follows:

Define the Deng's slope of the sequence $X_i(i=0,1,2,\cdots,m)$ at k is $Z_i(k)$, which is calculated as follows:

$$Z_i(k)=\frac{\min\limits_i\min\limits_k\left|\frac{x_i(k)-x_i(k-1)}{\frac{1}{m(n-1)}\sum\limits_{i=1}^{m}\sum\limits_{k=2}^{n}\left|x_i(k)-x_i(k-1)\right|}\right|+\rho\max\limits_i\max\limits_k\left|\frac{x_i(k)-x_i(k-1)}{\frac{1}{m(n-1)}\sum\limits_{i=1}^{m}\sum\limits_{k=2}^{n}\left|x_i(k)-x_i(k-1)\right|}\right|}{\left|\frac{x_i(k)-x_i(k-1)}{\frac{1}{m(n-1)}\sum\limits_{i=1}^{m}\sum\limits_{k=2}^{n}\left|x_i(k)-x_i(k-1)\right|}\right|+\rho\max\limits_i\max\limits_k\left|\frac{x_i(k)-x_i(k-1)}{\frac{1}{m(n-1)}\sum\limits_{i=1}^{m}\sum\limits_{k=2}^{n}\left|x_i(k)-x_i(k-1)\right|}\right|} \tag{12}$$

where $\rho\in[0,1]$,usually $\rho=0.5$.

The relational degree between X_i and X_0 is defined as:

$$\gamma(X_0,X_i)=\begin{cases}\dfrac{1}{n-1}\sum\limits_{k=2}^{n}\dfrac{\operatorname{sgn}((x_0(k)-x_0(k-1))\cdot(x_i(k)-x_i(k-1)))}{1+\frac{1}{2}\left|Z_0(k)-Z_i(k)\right|+\frac{1}{2}(1-\frac{\min(Z_0(k),Z_i(k))}{\max(Z_0(k),Z_i(k))})}\\ 1\end{cases} \tag{13}$$

When $(x_0(k)-x_0(k-1))$ and $(x_i(k)-x_i(k-1))$ are not 0 at the same time, $\gamma(X_0,X_i)$ equals the upper formula; otherwise $\gamma(X_0,X_i)$ equals 1.

And also,

$$\begin{aligned}&\operatorname{sgn}((x_0(k)-x_0(k-1))\cdot(x_i(k)-x_i(k-1)))\\&=\begin{cases}1 & when(x_0(k)-x_0(k-1))\cdot(x_i(k)-x_i(k-1))\geq 0\\ -1 & when(x_0(k)-x_0(k-1))\cdot(x_i(k)-x_i(k-1))<0\end{cases}\end{aligned} \tag{14}$$

5. Early Warning of Failure Probability

When the new meteorological condition is obtained, first, determine which classification it belongs to according to the fuzzy clustering, and then calculate the relational degree between the reference sequence and the comparison sequence among the classification, final, calculate the failure rate p under the meteorological condition according to weighted principle.

$$p = \frac{\sum_{i=1}^{n} p_i \gamma_{0i}}{\sum_{i=1}^{n} \gamma_{0i}} \tag{15}$$

Where γ_{0i} is the gray relational degree of the reference sequence and the i -th comparison sequence in the classification, and p_i is the failure rate of the i -th comparison sequence in the classification.

6. Example

Based on the historical data of a city power supply line, 10 sets of data sequences are selected for example verification. As shown in Table 1.

Table 1. The statistics of line failure rate

Meteorological sequence	Maximum wind speed (m/s)	average temperature (°C)	average relative humidity (%)	rainfall (mm)	Line failure rate (times·100/month/km)
x1	10.1	25.2	73	70.9	0.9414
x2	5.2	24.3	79	32.3	0.3096
x3	4.1	32.3	85	19.8	0.5616
x4	4.3	30.7	84	119.5	0.4968
x5	9.2	25	81	75.1	0.8316
x6	3.5	30.5	89	25.7	0.5292
x7	4.9	32.6	90	29.7	0.6156
x8	8.2	24.9	80	82	0.8946
x9	9	26.3	78	86.5	0.8496
x10	3.8	31.6	88	23.2	0.5868

Using the coefficient of variation method, through the formula (1)~(3), the weight value of the maximum wind speed, the average temperature, the average relative humidity and the rainfall is $a = [0.3406, 0.1006, 0.0543, 0.5045]$.

Theoretically, the rainfall and the maximum wind speed are likely to cause the line fault, while the temperature and humidity have less influence. Therefore, the weight value obtained by the coefficient of variation method is consistent with the actual situation, which meets the requirements of the rationality analysis.

IOP Conf. Series: Earth and Environmental Science **133** (2018) 012001 doi:10.1088/1755-1315/133/1/012001

The original data are normalized and weighted according to (4)~(8), and the transitive closure matrix is obtained by the square method. The transitive closure matrix $t(R)$ is obtained as follows.

$$t(R)=\begin{bmatrix} 1 & 0.2899 & 0.2899 & 0.4650 & 0.8543 & 0.2899 & 0.2899 & 0.8543 & 0.8543 & 0.2899 \\ 0.2899 & 1 & 0.4626 & 0.2899 & 0.2899 & 0.4626 & 0.4626 & 0.2899 & 0.2899 & 0.4620 \\ 0.2899 & 0.4626 & 1 & 0.2899 & 0.2899 & 0.7371 & 0.6099 & 0.2899 & 0.2899 & 0.7371 \\ 0.4650 & 0.2899 & 0.2899 & 1 & 0.4650 & 0.2899 & 0.2899 & 0.4650 & 0.4650 & 0.2899 \\ 0.8543 & 0.2899 & 0.2899 & 0.4650 & 1 & 0.2899 & 0.2899 & 0.8630 & 0.8630 & 0.2899 \\ 0.2899 & 0.4626 & 0.7371 & 0.2899 & 0.2899 & 1 & 0.6099 & 0.2899 & 0.2899 & 0.7585 \\ 0.2899 & 0.4626 & 0.6099 & 0.2899 & 0.2899 & 0.6099 & 1 & 0.2899 & 0.2899 & 0.6099 \\ 0.8543 & 0.2899 & 0.2899 & 0.4650 & 0.8630 & 0.2899 & 0.2899 & 1 & 0.8691 & 0.2899 \\ 0.8543 & 0.2899 & 0.2899 & 0.4650 & 0.8630 & 0.2899 & 0.2899 & 0.8691 & 1 & 0.2899 \\ 0.2899 & 0.4626 & 0.7371 & 0.2899 & 0.2899 & 0.7585 & 0.6099 & 0.2899 & 0.2899 & 1 \end{bmatrix}$$

The cross matrix λ, obtained by passing the closure matrix, is $\lambda=[1,0.8691,0.8630,0.8543,0.7585,0.7371,0.6099,0.4650,0.4626,0.2899]$

According to the cross matrix λ, the dynamic clustering diagram is shown in Fig.1

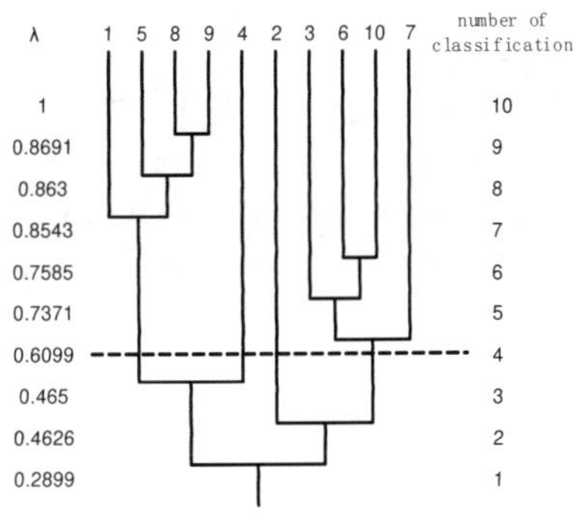

Figure 1. Dynamic clustering graph

This paper takes the significance level $\alpha=0.05$, by querying the F distribution table, according to the formula (9)~(11), calculates F and $(F-F_{\alpha})/F_{\alpha}$ in each classification. The results are shown in Table 2.

Table 2. Distribution table of F and $(F-F_{\alpha})/F_{\alpha}$

Classification number	7	6	5	4	3	2
F	24.266	37.594	53.186	63.200	12.167	24.277
F distribution table	8.94	6.26	5.19	4.76	4.74	5.32
ratio	1.714	5.005	9.248	12.277	1.567	3.563

The results show that the samples divided into four classes have the best results, as $0.465 < \lambda \le 0.6099$. According to the dynamic clustering graph, the specific classification is: the first classification $L_1 = \{x_1, x_5, x_8, x_9\}$, the second classification $L_2 = \{x_2\}$, the third classification $L_3 = \{x_3, x_6, x_7, x_{10}\}$, the fourth classification $L_4 = \{x_4\}$. Taking the sample x10 as an example, according to the formula (13)~(14), the not improved gray relational degree is $\gamma = [0.7752, 0.8116, 0.6435]$. According to the formula (15)~(17), the improved gray relational degree is $\gamma = [0.9189, 0.9102, 0.8076]$.

According to (18), predicted failure rate is: not improved gray relational degree is $p = 0.5654$; improved gray relational degree is $p = 0.5670$.

The predicted failure rate of all samples is shown in Table 3.

Table 3. Table of failure rate prediction

Gray relation		Not improved		Improved	
Sample numble	Line failure rate	Predicted failure rate	Difference percentage	Predicted failure rate	Difference percentage
x1	0.9414	0.8572	0.0894	0.8583	0.0883
x2	0.3096	0.3096	0.0000	0.3096	0.0000
x3	0.5616	0.5785	0.0301	0.5766	0.0267
x4	0.4968	0.4968	0.0000	0.4968	0.0000
x5	0.8316	0.8948	0.0760	0.8956	0.0770
x6	0.5292	0.5880	0.1111	0.5872	0.1096
x7	0.6156	0.5593	0.0915	0.5595	0.0911
x8	0.8946	0.8700	0.0275	0.8737	0.0234
x9	0.8496	0.8856	0.0424	0.8891	0.0465
x10	0.5868	0.5654	0.0365	0.5670	0.0337
Average difference			0.0504		0.0496

It can be concluded from the table that the improved gray relational analysis improves the accuracy of the prediction and the predicted difference is within the allowable range, which verifies the effectiveness of the proposed method.

References

[1] Dewen Wang and Zhiwei Sun 2015 Big data analysis and parallel load forecasting of electric power user side *Proceedings of the CSEE* vol 35 pp 527-537.

[2] Teng Zhao, Yan Zhang and Dongxia Zhang 2014 Application technology of big data in smart distribution grid and its prospect analysis *Power System Technology* vol 38 pp 3305-3312.

[3] Jie Gai, Qunzhan Li and Jia Wang 2016 Modal parameter identification of low frequency oscillation through NExT-ERA based on fuzzy clustering *Power System Protection and Control* vol 44 pp 40-49.

[4] Yongjun Zhang, Chao Chen and Liang Xu 2011 Prediction of original reliability parameters of power system based on fuzzy clustering and similarity *Power System Protection and Control* vol 39 pp 1-5.

[5] Guifen Chen, Liying Cao and Guowei Wang 2009 Application of weighted spatially fuzzy dynamic clustering algorithm in evalution of soilfertility *Scientia Agricultura Sinica* vol 42 pp 3559-3563.

[6] Guowei Wang, Li Yan and Guifen Chen 2010 Weighted spatially fuzzy dynamic clustering algorithm *Computer Engineering and Applications* vol 46 pp 146-149.

[7] Lianju Ning and Meng Li 2011 Evaluation model for large and medium-sized industrial enterprises' technological innovation capability based on factor analysis method *Science Research Management* vol 32 pp 51-58.

[8] Wei Zhao, Jian Lin and Shufang Wang 2013 Influence of human activities on groundwater environment based on coefficient variation method *Environmental Science* vol 34 pp 1277-1283.

[9] Ganyun Lv, Haozhong Chen and Haibao Zhai 2004 Fault diagnosis of power transformer based on improved grey relationl analysis *Proceedings of the CSEE* vol. 24; pp. 121-126.

[10] Jian Shen, Shujuan Shi and Yang Zhou 2014 Surface water environmental quality assessment of Danjiankou valley based on improved grey relational analysis *Environmental Monitoring in China* vol 30 pp 41-46.

A Novel Fault Elements Location Algorithm Based on Wide Area Fault Recorder Data for HV Transmission System

Juntao He, Renfei Che*, Xiaocheng Meng, Qingmeng Meng
Key Laboratory of Power System Intelligent Dispatch and Control
School of Electrical Engineering, Shandong University, Jinan, China

E-mail: 2366481139@qq.com, cherenfei@sdu.edu.com

Abstract. After a large-scale grid failure, it is important for dispatchers to know the faulty components and causes and to develop the correct scheduling strategy. In this paper, a new fault diagnosis scheme is presented based on widely configured fault recorders. The proceeded algorithm using the wide-area recorder data to locate the fault components such as transmission lines, busbars and transformers. Different indicators are given for different components on the grid. A variety of criterion indexes are deduced for the transmission line using the distributed parameter model. The credibility of the judgment is improved by using multiple criteria. A search scheme based on the initial adjacency matrix is used to locate fault component on the grid. The fault area is defined according to the uploaded recorder data. The fault components are searched in the wide defined fault area according to the proposed scheme. The element with the greatest probability of failure is found. The three faulty elements simulation models based on PSCAD/EMDTC are established and studied, what's more, the power system model is also established to test the algorithm. The simulation results verify the effectiveness of the proposed method.

1. Introduction

Large-scale transmission grid is the backbone of power systems, at the same time, transmission system failures are always inevitable. Thus, large power systems call for precise and effective fault diagnosis algorithms to improve the overall safety and reduce economic losses. To solve this problem, many methods of fault diagnosis, such as expert system [1]; Bayesian network [2]; Petri nets [3]; artificial neural networks (ANNs) [4]; optimization technology [5] are proposed.

Most diagnosis algorithms are protection system information or switch information based [6]. But this kind of algorithm is not reliable. For example, a mismatch between expected and real switch action could result in false diagnosis result. To solve the reliability problem in [6], the method Using several agents corresponding to several methods cooperate to fulfill fault diagnosis is proposed [7], but how to

IOP Conf. Series: Earth and Environmental Science **133** (2018) 012002 doi:10.1088/1755-1315/133/1/012002

properly combine different methods and how these agents cooperate are unsolved issues. With the development of phasor measurement units (PMU), the fault diagnosis model based on PMU data has also been studied [8]. But high cost of the PMU limits the wide configuration and thence popularization of this algorithm.

In order to help the dispatchers to locate fault components quickly after power grid fault, a fault diagnosis scheme based on fault recorder data is proposed in this paper. The fault criteria index of transmission line, transformer and bus are proposed in section2. Section 3 introduces the fault element search method based on the adjacency matrix. In section 4, We show the PSCAD/ EMTDC simulation results demonstrate the correctness of the proposed method.

2. Fault recognition index for different elements

2.1. Fault criteria of transmission line

2.1.1. Ratio criterion. The fault criteria using the synchronized voltage and current samples at both ends of a transmission line to calculate the location of the fault is presented in this section. Considering a transmission line with a non-fault distribution parameter model. Then both voltage and current measured at a distance l km away from receiving end obey the partial differential equations:

$$\begin{bmatrix} \dot{U}_s \\ \dot{I}_s \end{bmatrix} = \begin{bmatrix} \cosh(\gamma l) & Z_c \sinh(\gamma l) \\ \sinh(\gamma l)/Z_c & \cosh(\gamma l) \end{bmatrix} \begin{bmatrix} \dot{U}_r \\ \dot{I}_r \end{bmatrix} \tag{1}$$

Where, γ is called propagation constant of the transmission line respectively.

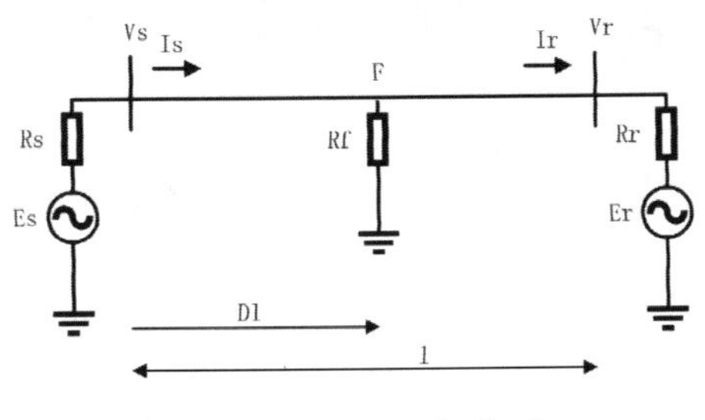

Figure 1. A single-phase faulted transmission line

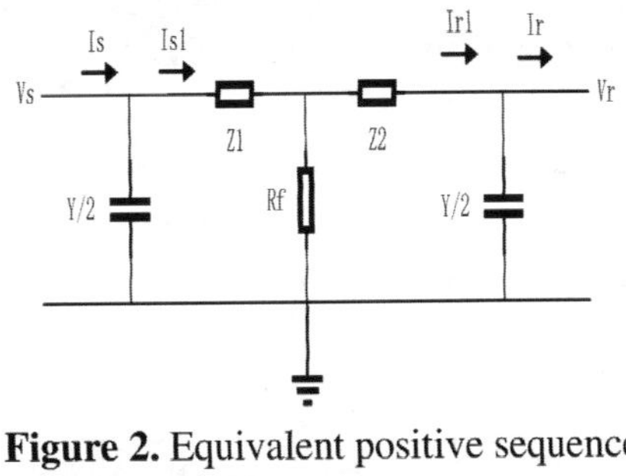

Figure 2. Equivalent positive sequence network of a fault line

Figure 1 shows the schematic diagram of transmission line with fault. When an internal fault occurs at the point F (Dl km away from sending end), voltage at F can be expressed as:

$$\dot{V}_F = \dot{U}_s \cosh(\gamma Dl) - \dot{I}_s Z_c \sinh(\gamma Dl) \tag{2}$$

$$\dot{V}_F = \dot{U}_r \cosh(\gamma (1-D)l) + \dot{I}_s Z_c \sinh(\gamma (1-D)l) \tag{3}$$

(2) and (3) represent the post-fault voltages at fault point expressed in terms of measured sending and receiving data pair (VS; IS) and (VR; IR), respectively. From (2) and (3), the following expressions can be derived:

$$V_s = V_r + I_s Z_c \sinh(\gamma Dl) - V_s(\cosh(\gamma Dl) - 1) + I_r Z_c \sinh(\gamma(1-D)l) - Vr(\cosh(\gamma(1-D)l) - 1) \quad (4)$$

$$T * \sinh(\gamma Dl) - J * \cosh(\gamma Dl) = 0 \quad (5)$$

T is the index of fault recognition. When the fault occurs external or no fault occurs, Dl is zero, so the absolute value of T and J are zero. Since such measured components all satisfy the transmission line equation, the computed absolute values of T and J will all be held at zero before the occurrence of a fault. This result can be derived using mathematical formulas for T.

When internal fault occurs, $\cosh(\gamma Dl)$ and $\sinh(\gamma Dl)$ in equation (5) are non-zero values and so does the index T. Thus, the different index value before and after the fault can be used as criteria index of the fault detection of transmission line.

$$|T| = |V_s - V_r \cosh(\gamma l)) - I_r Z_c \sinh(\gamma l)| \quad (6)$$

In order to distinguish fault, the T index before and after the fault are both calculated and used λ as the fault discrimination index of transmission line. When internal fault occurs, the calculated value is very large. A threshold variant λ is introduced whose expression is shown in equation (7).

$$\lambda = |T^{[1]} / T^{[0]}| \quad (7)$$

When the calculated value is greater than the threshold value, the line is regarded as faulty component. It should be noted that no assumptions in the procedure of derivation for the fault detection/location index λ are made. Hence, the index λ is hardly affected by the variations of source impedance, load change, fault impedance, fault inception angle and fault type.

2.1.2. Phase criteria index. Consider of an equivalent positive sequence network of a faulted line shown in Figure 2. It will be shown that this network is enough to detect the location of the fault. While the parameters of the line are represented by the distributed parameter model. The following equations can be obtained:

$$\dot{I}_{s1} = \dot{I}_s - \dot{V}_s * Y' / 2 \quad (8)$$

$$\dot{I}_{r1} = \dot{I}_r + \dot{V}_r * Y' / 2 \quad (9)$$

Fault direction P could be defined from equation (1) and (2) based on the current direction.

$$P = \left|\arg\left(\dot{I}_s - \dot{V}_s * Y' / 2\right) / (\dot{I}_r + \dot{V}_r * Y' / 2)\right| \quad (10)$$

In theory, when a fault occurs in the internal area, the value of P is π. And when there is no internal fault the value of P comes to zero. For internal fault cases, the absolute value of the rises to a certain value with a large slope in practice. For external fault or none-fault cases, the absolute value of the will fast decay to zero. For the sake of high reliability, we set a threshold incorporated with a counter limit to recognize the faulted zone.

2.1.3 Differential current criteria index. For a normally operating two-terminal transmission line, the sum of the current phasors at both ends is zero. However, when fault occurs, the result of post-fault double-ended current phasors will be equal to the fault current value. Therefore, differential current can be calculated as a criterion for fault detection.

$$\dot{I}_{\varphi f} = \left|\dot{I}_{\varphi m} + \dot{I}_{\varphi n}\right| \tag{11}$$

Where φ represents current phase which can be a, b or c.

2.2. *Fault indicators for transformer and bus*

2.2.1. Fault indicator for transformer. Differential current protection is commonly used as a main protection of the standard and non-standard power transformers. The calculation of the differential current at both ends of the transformer is rapid and sensitive to internal fault of the transformer.

Differential current is measured by adding the vector current entering and leaving the circuit. The differential current measuring principle (DCMP) reflects the kind of circuit it reflects. Thus, In this paper, the differential current is adopted as the fault detection index of power transformer.

For the ubiquitous YD11-type transformer, it is necessary to consider the phase shift of the voltage and current. The positive sequence voltage and current of the star connect side lag 30 degrees behind the phasors of the delta connect side respectively. So we can get the differential current index for phase shift transformer：

$$\dot{I}_{r\varphi} = e^{j*pi/6}\dot{I}_{Y\varphi} * N + \dot{I}_{\Delta\varphi} \tag{12}$$

2.2.2. Fault indicator for bus. Like transformers, each outgoing line's current on the bus satisfies the KCL law. And in this paper, we also take the differential current as the fault detection criteria of bus:

$$\dot{I}_r = \left|\dot{I}_1 + \dot{I}_2 + \cdots + \dot{I}_n\right| \tag{13}$$

Where $\dot{I}_k$ represents the current of line k connected with the bus (k is 1, 2, 3…n).

3. Fault element search scheme

Regarding power grid as a graph G without directions, and graph theory is used to simplify the complex electric network. The bus is denoted as vertex (V(G)), and transformer and transmission line is denoted as edges(E(G)). The connection between two nodes is expressed as a function(f(G)). The topology of the power system can be described as G=(V(G), E(G), f(G)). The adjacency matrix $A = \sum a_{ij}$ is used to show the connection of the graph. The elements (a_{ij}) of the matrix represent the connection between vertex i and j. If vertex i and vertex j are directly connected, then $a_{ij} = 1$, otherwise $a_{ij} = 0$. If a_{ij} is not zero, record the branch connected with the two vertexes.

When fault occurs, the fault area is firstly ascertained using data from fault recorder. Then the adjacency matrix of the faulty area A_{fault} is formed based on fault recorder's location. According to adjacency matrix A_{fault}, component searches are performed. The search starts with the boundary node of the adjacency matrix, ending with the faulty element or the last node, making sure that each component is checked.

4. Simulation analysis

4.1. Performance of line fault index. To examine the performance of the proposed index, a model of a double-ended power supply with the rated voltage of 500 kV and the line length of 360 km, which is established using the PSCAD/EMDTC simulator, is adopted to evaluate the performance of the proposed scheme. The sampling frequency is 5000 Hz. Single phase grounding fault (take phase a for example) is simulated in internal and external of the line separately, and figures as follows show the absolute value of various indicators before and after the fault. Besides, all of the indexes are calculated based on positive sequence parameters and the sampling frequency is 5000 Hz. Figure 3 shows the ratio and T index curves, and the radio index of the post-fault is far greater than the pre-fault obviously. The fault ratio indexes of internal and external fault which are represented by positive sequence are showed in Figure 4. We can see that the ratio indexes between pre-fault and post-fault are very different. So, the ratio indicator can be used to distinguish the faults inside and outside the zone. Like the ratio indicator, Figure 5 and Figure 6 show the performance of the P indicator in detecting faults and distinguishing external and internal faults respectively. The curves of the P indicator are compared in these figures prove that it has good performance in fault detection.

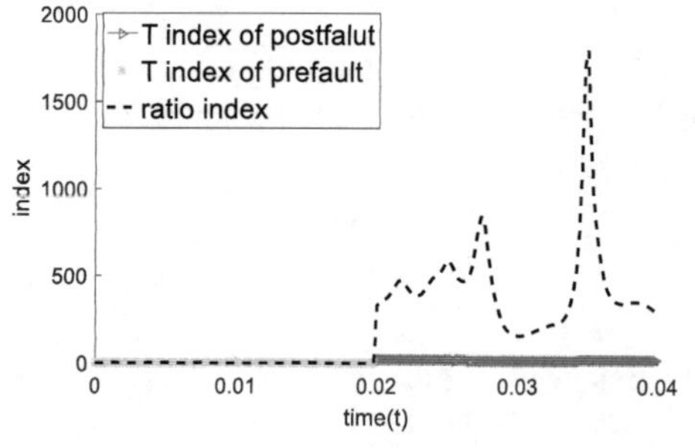

Figure 3. The behaviors of fault ratio index and T index under positive sequence for a-g fault

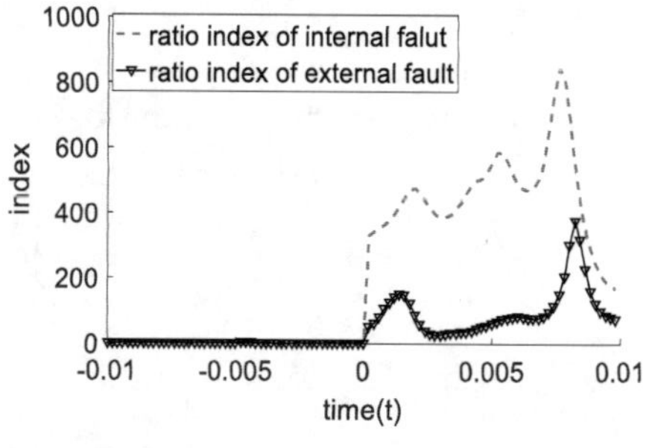

Figure 4. The behaviors of fault ratio index under positive sequence between internal and external falut

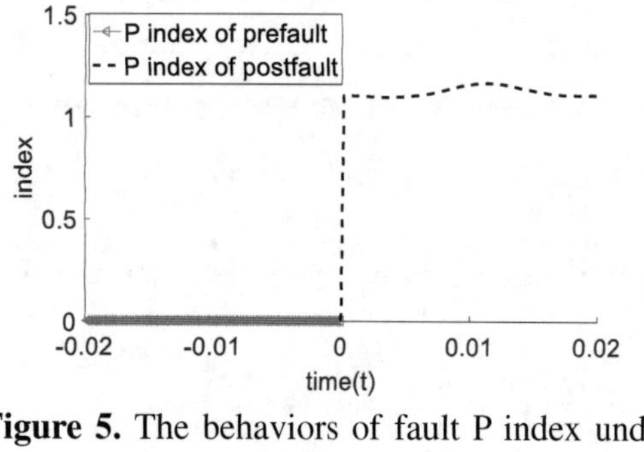

Figure 5. The behaviors of fault P index under positive sequence for a-g fault

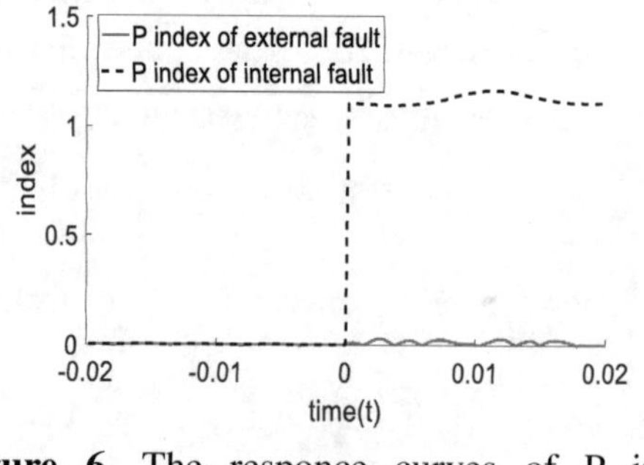

Figure 6. The responce curves of P index between internal and external fault

4.2 Fault component location. In order to test the performance of the fault locating algorithm in the power grid, the proposed method is applied to the IEEE 9-bus test system. Table 1 shows the results of various indicators after the fault occurs and gives the faulty components located. Simulation results show that the proposed algorithm can quickly search for faulty components and locate them.

Table 1. The results of fault components location in the 9-bus test system

Fault component	Fault type	Fault discrimination			Result
		Ratio index	P index	Differential current	
Line 1	AG	442.485	1.4745	2.5656	Line 1
	BC	860.824	2.4873	2.8031	
	BCG	1040.4	2.6067	3.0147	
	ABC	1606.5	2.7993	3.2396	
T 3	AG			1.9357	T3
Bus 5	AG			1.8952	Bus 5

5. Conclusion

A fault element location algorithm for power system based on wide area fault recorder data is presented in this paper. Different fault detection indicators of power grid components are proposed, and elements searching method is given, too. After the fault occurs, the fault area and components can be quickly identified, which plays an important role in the safe and stable operation of the power grid. Simulation cases also verifies the effectiveness of the algorithm.

References

[1] Fukui C and Kawakami J 1986 An expert system for fault section estimation using information from protective relays and circuit breakers *IEEE Trans. on Power Delivery*. **1** 83-91.

[2] Zhu Yongli, Huo Limin, and Lu Jinling 2006 Bayesian networks-based approach for power systems fault diagnosis *IEEE Transactions on Power Delivery*. **21** 634-639.

[3] Yang Jianwei, and He Zhengyou 2005 Power system diagnosis approach based on time sequence fuzzy Petri net *Automation of Electric Systems*, **35** 46-51.

[4] Thukaram D, Khincha P and Vijaynarasimha H P 2005 Artificial neural network and support vector machine approach for locating faults in radial distribution systems *IEEE Transactions on Power Delivery* **20** 710-21.

[5] Wen Fushuan and Han Zhengxiang 1994 Fault section estimation in power systems using genetic algorithm and simulated annealing. *Proceedings of the CSEE* pp 1429-35.

[6] Weng Hanlin, Mao Peng, Lin Xiangning， et al 2007 An improved model for optimizing power fault diagnosis *Automation of Electric Systems*. **31** 66-70.

[7] S. Lin, Z. Y. He, Q and Q. 2010 Qian Review and development on fault diagnosis in power grid *Power Syst. Protect. Control* **38** 140-150.

[8] P. Z. Dehkordi, A. S. Dobakhshari and A. M. Ranjbar 2012 Artificial neural network and support vector Machine approach for locating faults in radial distribution systems *3rd IEEE PES Innovative Smart Grid Technologies Europe* (Berlin: ISGT Europe) pp 1-9.

IOP Conf. Series: Earth and Environmental Science **133** (2018) 012003 doi:10.1088/1755-1315/133/1/012003

Magnetic field simulation and shimming analysis of 3.0T superconducting MRI system

Z. K. Yue, Z. Z. Liu, G. S. Tang, X. C. Zhang, L. J. Duan, and W. C. Liu
School of Electrical Engineering, Shandong University, Jinan 250061, Shandong Province, China

yuezengkai2016@163.com, liuzhizhen@sdu.edu.cn

Abstract. 3.0T superconducting magnetic resonance imaging (MRI) system has become the mainstream of modern clinical MRI system because of its high field intensity and high degree of uniformity and stability. It has broad prospects in scientific research and other fields. We analyze the principle of magnet designing in this paper. We also perform the magnetic field simulation and shimming analysis of the first 3.0T/850 superconducting MRI system in the world using the Ansoft Maxwell simulation software. We guide the production and optimization of the prototype based on the results of simulation analysis. Thus the magnetic field strength, magnetic field uniformity and magnetic field stability of the prototype is guided to achieve the expected target.

1. Introduction

In recent years, MRI has become an important non-invasive tomography technique, which plays an increasingly important role in clinical diagnosis and treatment. High field MRI system has a great advantage in SNR, resolution and scanning time. So the detection rate of microstructure and minimal lesion of high field MRI system is significantly better than that in middle and low MRI system. It can also shorten the patient's examination time and carry out the research of spectrum and functional imaging. And high field MRI system has become the mainstream of the current market [1].

The first cross-sectional magnetic resonance images of human fingers were taken by Mansfield and Maudsley in 1973 and they shared the 2003 Nobel Prize in Physiology or Medicine [2]. The first commercial magnetic resonance scanner in Europe was built in 1983 at the Radiology Diagnostics School at Manchester Medical University. The magnetic field strength of most MRI systems on the market was less than or equal to 0.6T since they were applied to medical practice in the 1970s. In China, the application of superconducting technology began in the 1960s and 1970s. The first 1.5T MRI system appeared in the early 1980s and became the dominant magnetic field strength in clinical high-quality magnetic resonance imaging systems in the late 1980s. In the 80s and 90s, China Kejian Co. Ltd developed the first 0.5T superconducting magnetic resonance magnet [3-6].

The 3.0T/850 superconducting MRI molecular imaging system meets the high quality imaging requirements. The magnetic induction intensity of its magnet center is stronger or equal to 3.0T. The magnetic field uniformity of its 45cm spherical imaging region after shimming is less than 10ppm, of which the magnetic field stability is less than 0.1ppm/h. And its effective magnetic field keeps longer, which can reduce maintenance costs and maintenance frequency and ensure the definition of imaging. It can conduct research on spectroscopy, molecular imaging and functional imaging because it has obvious advantages in SNR and DPI. The 3.0T/850 superconducting MRI molecular imaging system is an advanced dual platform of clinical and research [7].

2. Theory

In this paper, we mainly studied designing principles of superconducting magnet and principles of magnetic field shimming.

2.1. Principle of magnet designing

In this paper, the superconducting magnet is made of a number of hollow cylindrical coils, which is designed according to the principle of the sixth-order coil designing [8]. The magnetic field generated by the superconducting magnet is treated as the superposition of magnetic fields generated by several hollow cylindrical coils. Fig. 1 shows cross section diagram of the hollow cylindrical coil. The current in the superconductor is evenly distributed when the magnetic field strength is much higher than the critical magnetic field of the superconductor. The formula of field strength distribution of hollow cylindrical coil with uniform current distribution is shown in (1). Among them, j is the current density（A/cm^2） and λ is the filling factor. The materials of superconducting wire are used in the commonly used niobium titanium alloy.

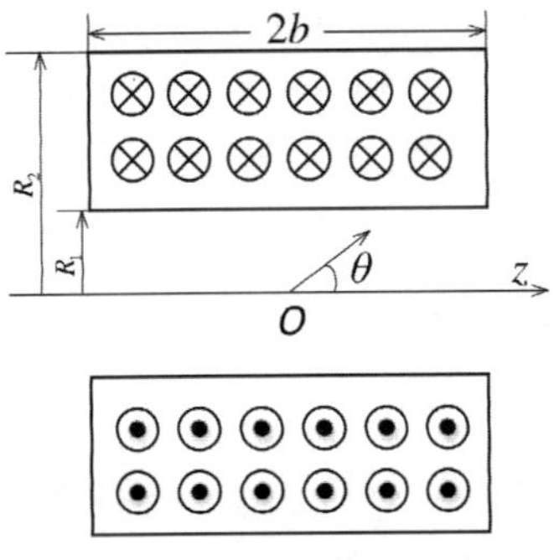

Figure 1. Simplified schematic diagram of hollow cylindrical coil.

$$\begin{aligned}
&B_n(j,\theta)=\frac{2\pi}{5}j\lambda R_2\left[M_0+M_2(\frac{r}{R_2})^2P_2+M_4(\frac{r}{R_2})^2P_4+M_6(\frac{r}{R_2})^2P_6+\dots\right]\\
&\beta=\frac{b}{R_1},\alpha=\frac{R_2}{R_1}\\
&C_1=\frac{1}{1+\beta^2},C_2=\frac{\beta^2}{1+\beta^2},C_3=\frac{\alpha^2}{\alpha^2+\beta^2},C_3=\frac{\beta^2}{\alpha^2+\beta^2}\\
&M_0=\beta/\alpha\ln\frac{\alpha+\sqrt{\alpha^2+\beta^2}}{1+\sqrt{1+\beta^2}},M_2=\frac{1}{2}\alpha/\beta(C_1^{3/2}-C_3^{3/2})\\
&M_4=\frac{1}{12}\alpha^3/\beta^3\left[C_1^{3/2}(1+\frac{3}{2}C_2+\frac{15}{2}C_2^2)-C_3^{3/2}(1+\frac{3}{2}C_4+\frac{15}{2}C_4^2)\right]\\
&M_6=\frac{1}{30}\alpha^5/\beta^5\left[C_1^{3/2}(1+\frac{3}{2}C_2+\frac{15}{2}C_2^2-\frac{35}{4}C_2^3+\frac{315}{8}C_2^4)-C_3^{3/2}(1+\frac{3}{2}C_4+\frac{15}{8}C_4^2-\frac{35}{4}C_4^3+\frac{315}{8}C_4^4)\right]\\
&\cos\theta=u,P_2=\frac{1}{2}(3u^2-1),P_4=\frac{1}{8}(35u^4-30u^2+3)\\
&P_6=\frac{1}{16}(231u^6-315u^4+105u^2-5)
\end{aligned}\quad(1)$$

2.2. Magnetic field simulation and shimming analysis

The design of the main magnet system requires a highly uniform and highly stable magnetic field. It must be solved numerically according to the theory of electromagnetic field with reliable and convenient computer software. We use Ansoft Maxwell simulation software to simulate the magnetic field distribution of the superconducting magnet in order to provide a scientific basis for the design and optimization of the magnet.

The magnetic field of superconducting magnet can be influenced by its design, manufacturing, installation environment and other factors. Thus the uniformity of the main magnetic field needs to be adjusted to achieve a certain degree to obtain high quality magnetic resonance images. The non-source shimming technology distribute the magnetic field evenly by optimizing the distribution of a series of shimming pieces (magnetic material, usually electrical steel) [9]. In practice, the number of shimming pieces should be controlled at a minimum when shimming requirements are met. Because the presence of shimming pieces may introduce thermal stability and eddy current problems. In this paper, we calculate the shimming radius using the extreme location method to regulate the position of shimming pieces. To avoid repeated correction, successive approximation is presented to calculate the number of shimming pieces [10].

3. Simulation and experimental results

The simulation model of this superconducting magnet is established using Ansoft Maxwell simulation software. We get the distribution curve of magnetic strength along axial direction of the magnet (as shown in Fig. 2) and the magnetic field distribution in the 45cm spherical imaging region (as shown in Fig. 3).

IOP Conf. Series: Earth and Environmental Science **133** (2018) 012003 doi:10.1088/1755-1315/133/1/012003

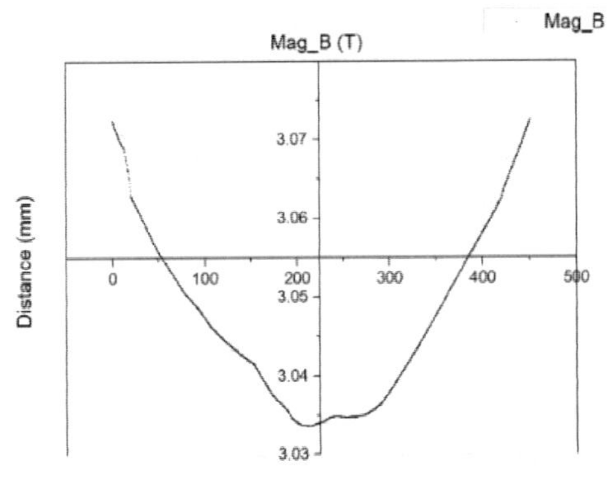

Figure 2. Distribution curve of magnetic strength along axial direction.

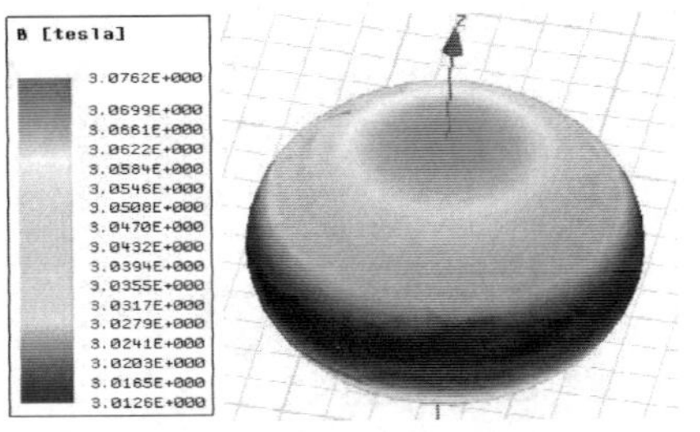

Figure 3. Magnetic field distribution in the 45cm spherical imaging region.

We add shimming pieces in the original magnet model through the analysis of simulation results. Then we calculate and optimize the location and quantity of shimming pieces. We also get the distribution curve of magnetic strength along axial direction (as shown in Fig. 4) and the magnetic field distribution (as shown in Fig. 5) after shimming.

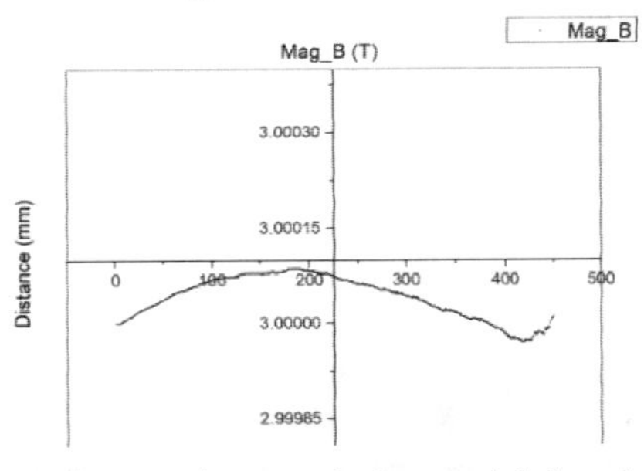

Figure 4. Distribution curve of magnetic strength along axial direction after shimming.

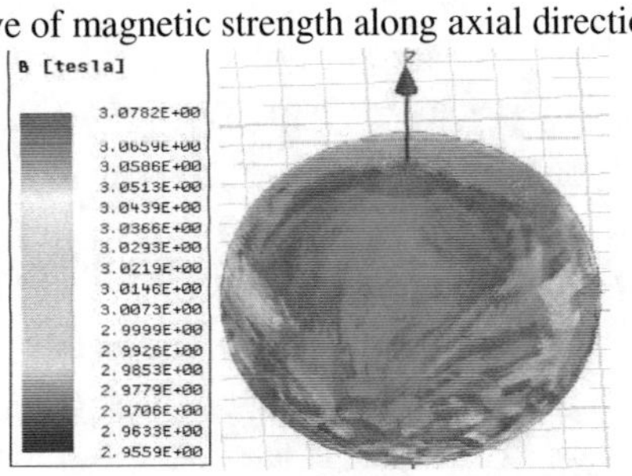

Figure 5. Distribution curve of magnetic strength along axial direction after shimming.

According to the simulation results, the prototype is designed and optimized. The superconducting magnet of medical 3.0T/850 MRI system is excited successfully by Weifang Xinli Superconducting Magnetic Technology Co., Ltd on June 15, 2017. Then we perform a series of experiments. The magnetic field strength, magnetic field uniformity and magnetic field stability of the prototype achieve the expected target.

4. Conclusions

In this paper, we analyze the principle of magnet designing. In order to provide a scientific basis for the design and optimization of magnets. We perform the magnetic field simulation and shimming analysis of the 3.0T/850 superconducting MRI system using the Ansoft Maxwell simulation software. The prototype is designed and optimized according to the simulation results. Through testing the prototype as shown in Fig. 6, the magnetic induction intensity of its magnet center is stronger or equal to 3.0T. The magnetic field uniformity of its 45cm spherical imaging region after shimming is less than 10ppm, of which the magnetic field stability is less than 0.1ppm/h. Next, we will further optimize the parameters of the prototype.

Figure 6. Test results.

Acknowledgments

This research has received funding from the key technology innovation project of key industries in Shandong Province (2016JMRH0211). We cooperated with Weifang Xinli Superconducting Magnetic Technology Co., Ltd in Shandong Province

5. References

[1] Zhang X Y J 2008 The effect of magnetic field homogenous level on the quality of MRI *Biomedical Engineering & Clinical Medicine* 12(3) 245-247

[2] P. Mansfied and A. A. Maudsley 2014 “Medical Imaging by NMR” *Br J Radiol* 50(592) 188-194

[3] K R Shao, S L Huang, J D Lavers, et al 2003 Analysis and optimization of a micro SMES system using multiple parallel solenoids IEEE Trans on Magnetics 39(5) 2615-2617

[4] B R Brandstatter Wolfgang Ring 1997 Optimization of a SMES device under nonlinear constraints Applied Computational Electromagnetics Society Journal 12(2) 50-53

[5] So Noguchi, Hideo Yamashita, Ishiyama 2003 An optimization method for design of SMES coils using YBCO tape IEEE Tran on Applied Superconductivity 13(2) 1856-1859

[6] P Alotto, B Brandstaetter, G Fuentratt, et al 1998 Some results on a SMES device optimization benchmark problem Int J of Applied Electromagnetics and Mechanics 9(3) 315-324

[7] Trattnig S, Friedrich K, Bogner W, et al M 2012 Advanced Musculoskeletal Magnetic Resonance Imaging at Ultra-high Field (7T) High-Field MR Imaging *Springer Berlin Heidelberg* 189-213

[8] Rhodes R G. J 1974 Superconducting magnet systems H. Brechna *Cryogenics* 14(6) 357-357

[9] Nd C J, Kelly W M J 1988 Common artifacts encountered in magnetic resonance imaging *Radiologic Clinics of North America* 26(5) 893

[10] Strite Podol'Skii A J 2000. Design procedure for permanent magnet assemblies with uniform magnetic fields for MRI devices *IEEE Transactions on Magnetics* 36(2) 484-486

IOP Conf. Series: Earth and Environmental Science **133** (2018) 012004 doi:10.1088/1755-1315/133/1/012004

HVDC Transmission an Outlook and Significance for Pakistani Power Sector

Muhammad Ahmad[1], Wang Zhixin[1], Wang Jinjian[1], Mazhar H. Baloach[2] Bao Longxin[3] and Qing Hua[4]

[1]Department of Electrical Engineering, Shanghai Jiao Tong University, 200240 Shanghai, China
[2]Department of Electrical Engineering, Mehran UET SZAB Khp Mir's Campus Khairpur, 66020, Sindh, Pakistan
[3]Shanghai Xilong Technology Co., Ltd, 201517 Shanghai, China
[4]Shanghai Huacheng Elevator Technology Co., Ltd, 201111 Shanghai, China

E-mail: wangzxin@sjtu.edu.cn

Abstract: Recently a paradigm shift in the power sector is observed, i.e., countries across the globe have deviated their attention to distributed generation rather than conventional centralized bulk generation. Owing to the above narrative, distributed energy resources e.g., wind and PV have gained the adequate attention of governments and researchers courtesy to their eco-friendly nature. On the contrary, the increased infiltration of distributed generation to the power system has introduced many technical and economical glitches such as long-distance transmission, transmission lines efficiency, control capability and cost etc. To mitigate these complications, the utility of high voltage direct current (HVDC) transmission has emerged as a possible solution. In this context, this paper includes a brief discussion on the fundamentals HVDC and its significance in Pakistani power sector. Furthermore, the potential of distributed energy resources for Pakistan is also the subject matter of this paper, so that significance of HVDC transmission can effectively be deliberated.

1. Introduction

To mitigate the destructive repercussions of carbon-containing fuel [1], eco-friendly distributed energy resources (DERs) have gained adequate importance these days. DERs can be regarded as more flexible power resources, albeit with limited capacity as compared to centralized power units. One of the biggest challenge to exploit the potential of DERs is the long distance bulk power transmission. Because the availability of these resources is far away from the load centres. Be contingent on the terminal voltage and current waveforms at DC side of HVDC, the HVDC topologies can be classified into two major categories; current source converters (CSC-HVDC) or Line-commutated converters (LCC) and voltage source converter (VSC-HVDC). Analytical models of 2-level VSC-HVDC, with fundamental circuit configuration, can be found in [2] and [8]. Comparative analysis of both VSC and CSC configurations is explicated in [3-4]. Recently developed modular multilevel converters (MMC) built with a cascaded structure of sub-modules, and in comparison with conventional two or three level MMC own several superior characteristics, e.g., low harmonics, high power quality and reduced AC-filter requirements, full details are outlined in [6]. Some applications of HVDC in different countries are argued in [6-8], whereas different projects of HVDC in China utilizing CSC and VSC configurations are listed in Table 1. Furthermore, in this paper; the current scenario of Pakistani power sector in the light of statistics, the potential of distributed renewable energy resources in Pakistan and recommendations to overcome the power shortage are the subject matters of this paper.

Table 1. The major applications of HVDC (including LCC based and VSC based) in China

Project	Year	Converter Type	Voltage Level	Capacity	Terminals
Gezhouba-Shanghai HVDC	1986	LCC	±500kV	1200MW	2
Sanxia-Changzhou HVDC	2003	LCC	±500kV	3000MW	2
Nanhui	2011	VSC	±30kV	20MW	2
Nan'ao	2013	VSC	±160kV	200MW	3
Zhousan	2014	VSC	±200kV	1000MW	5
Xiamen	2015	VSC	±320kV	1000MW	2

2. Pakistani Power Sector: An Outlook

Pakistan is one of the countries on the globe which have been suffering due to consistent power shortage since the beginning of 21st century. Power shortage in Pakistan has badly dented the economy and day to day affairs of common public. Subsequent governments have failed to address this critical issue. One biggest dilemma is that, although, the county is rich in renewable energy resources, because of the topographical location, yet the governments have failed to effectively utilize these resources to counter the power shortage. As it is an evident veracity that, the availability of DERs, i.e., wind and PV reside far away from the load centres. Thus, the radical changes in the power system transmission are required, so that power generated by DERs can efficiently be transmitted to the load centres or nearby grids for the purpose of utility.

2.1. Power Demand and Supply Statistical Analysis

This section of the paper briefly enlightens the statistical information of power demand and supply for last few years. Be reliant on these numbers, power supply and demand is forecasted till 2020 by the national power regulatory agency in Pakistan, which is called National Electric Power Regularity Authority (NEPRA). The facts reported in this section are extracted from State of Energy Report by NEPRA [9].

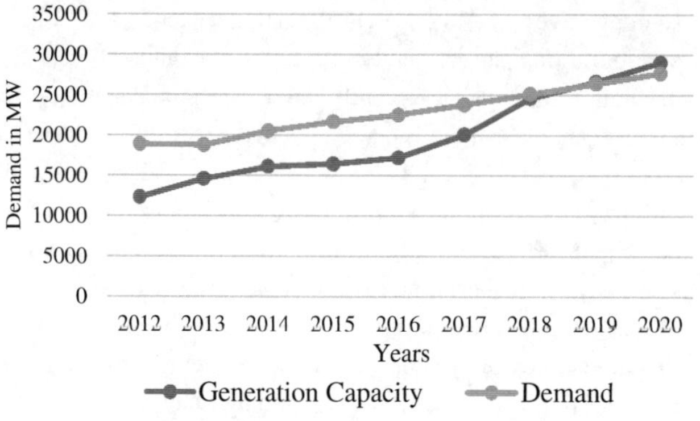

Figure 1. Demand & Supply Trends for Pakistani Power Sector [9]

From Figure 1, it can be observed that in recent years the maximum power deficit observed was approximately 6600 MW. If the supplementary generation of around 6000 MW is injected into the power system, the shortfall can approach to zero. Moreover, it should also be noted that in Figure 1 the data of K-Electric (private company) is not included, which is responsible for generation and distribution of power for the city of Karachi (capital of Sind province).

2.2. Potential of PV and Wind

Table 2. Solar Radiation Intensity & Wind Energy Assessment [10-11]

Region	Solar Radiation Intensity		Wind Energy Assessment		
	Minimum Intensity (W/m^2)	Highest Intensity (W/m^2)	Land Area (km^2)	Windy Area % of Total Land	Installable Capacity (MW)
Province Baluchistan	135.73	329.05	11709	15.71	58545
Province Sindh	145.29	331.27	29229	8.41	146145

Pakistan is one of the courtiers enriched in solar (PV) energy, the annual solar radiations for the different region of the country were calculated and explained comprehensively in the [10] and [12]. Be reliant on the important stats of solar radiation intensity discussed in [10], it can be concluded that solar radiation intensity remains suitable in different parts of the country from March to October. Similarly, Pakistan also enjoys the abundance of wind energy in the different parts of the county. The comprehensive analysis of wind potential in the different parts of the country is elaborated in [11]. Besides, recent developments in wind energy sector are elucidated in [13].

2.3. PV and Wind Generation Current Scenario

During FY-2016 solar power of 300 MW was added to the system. Therefore, the total power produced by solar energy has reached to 400 MW. Likewise, 50 MW of wind energy was added to the system, which makes the total of 306 MW. Finally, 63 MW from another renewable energy source i.e., bagasse was also added to the system and makes the total of 146 MW. Thus, the combined generation of all renewable energy resources at the end of the start FY-2017 was 852 MW [9].

3. VSC-HVDC: The Solution for Pakistani Power Sector

In order to connect the different regions of the country, to fully utilize the available energy resources long-distance power transmission power transmission is essentially required. In this regard HVDC projects in joint collaboration with the Chinese government is in progress. The project will utilize the potential of coal available in Thar (southeast part of the country), and then will be connected to the load centres situated in the eastern part of the country. Existing power transmission network of NTDC and route of the ongoing project are elucidated in Figure 2 and Figure 3.

Likewise, the northern regions of Pakistan are rich in hydroelectric potential, but far away from the load center. In addition, the southeast part of the country (e.g., Thatta in Sindh) is rich in wind energy, Whereas the substantial demand of load exists in the province of Punjab, which is the eastern part of the country and in the south part of the country (i.e., Karachi), which undeniably is the economic hub for Pakistan. So, to connect different parts of country economic and efficient network of the long-distance power transmission lines are becoming a necessity. Moreover, in future Government is planning to import electricity from the neighbouring counties e.g., Iran, China, India and Tajikistan, HVDC transmission can also be utilized for such purposes. Since, Pakistan's pioneer HVDC project is under development, in this regard Pakistan can seek guidance from the experience of other countries [3-4], where HVDC is being practised for a long time for different purposes. Depending upon the capacity of DC transmission circuit, Pakistan can adopt the VSC-HVDC or LCC-HVDC based topologies for long transmission. The details of ongoing HVDC project in Pakistan are elucidated in Table 3. Pakistan can seek guidance from the countries where HVDC is practised for a long time, and for the particular project mentioned in Table 3, LCC-HVDC can be a most pertinent option, due to the fact that, for higher capacity LCC-HVDC is practiced widely, and it can also be observed from the Chinese HVDC projects enlisted in Table 1.

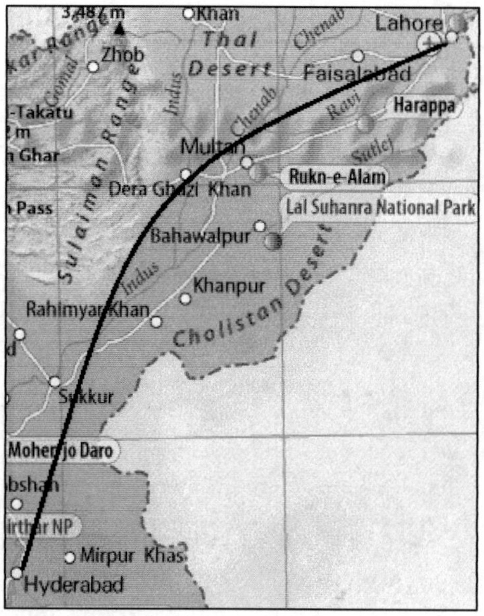

Figure 2: Map Indicating the Planned Route of Matiari to Lahore HVDC Transmission Line

Table 3. Minor Details of Ongoing HVDC Transmission Line Project in Pakistan [15]

Project	Voltage	Capacity	Distance	Company	Status
Matiari to Lahore HVDC Transmission Line	±660kV	4000MW	878 km	SGCC	Under Progress

As motioned earlier, in Pakistan renewable energy resource i.e., wind and PV reside far away from the load centres, and to completely utilize these resources long-distance power transmission lines are becoming an indispensable necessity. For that purpose, HVDC transmission has emerged as possible solution, and own several benefits over conventional AC transmission. The comprehensive analysis comprises of economic and technical comparison between HVAC and HVDC is debated in [16] and it was determined that for long distance power transmission HVDC is more suitable than HVAC.

4. Conclusion

A superficial view of Pakistani power sector was presented in this paper, which suggested that power shortage during peak hours was around 6600 MW. Moreover, it was observed that at the beginning of FY-2017 total generation by renewable energy resources was 852 MW, which surely is very less than the potential that Pakistan owns. Besides, the total transmission and distribution losses recorded by NTDC were approximately 18%. Thus, it can be concluded that the reduction of losses and increased utilization of renewable energy resources can unquestionably shrink the power shortage. To achieve these objectives, HVDC power transmission can prove pivotal for the power sector in Pakistan.

Acknowledgment

The authors would like to acknowledge the financing support of the National Natural Science Foundation of China (51377105), the Shanghai union plans (LM201702) and Science and Technology Program of Minhang District， Shanghai (2017MH271).

The Ongoing Challenge of Managing Carbon Monoxide

References

[1] National Research Council 2002
Pollution in Fairbanks, Alaska (The National Academics Press).

[2] Muhammad H. Rashid 2017 *Power Electronics Handbook* (Butterworth-Heinemann).

[3] Markus Christian Imhof 2015 *Voltage Source Converter Based HVDC-Modelling and Coordinated Control to Enhance Power System Stability* PhD Dissertation by ETH Zurich

[4] Dragan Jovcic and Khaled Ahmed 2015 *High Voltage DC Transmission Converter, Systems & AC-Grids* (John Willey)

[5] Michael P. Bahrman, Jan G, Johanss and Bo A. Nilsson 2003 Voltage Source Converter Transmission Technologies-The Right Fit for The Application *Power Engineering Society General Meeting* **3** pp 1847-1855

[6] Feng Wang, Lina Bertlig and Tuan Le 2011 An Overview Introduction of VSC-HVDC: State-of-art and Potential Applications in Electric Power Systems *Cigrè Int. Symp, Bologna* pp 1-9

[7] J. Glasdam, J. Hjerrild, Ł. H. Kocewiak and C. L. Bak 2012 Review on multi-level voltage source converter based HVDC technologies for grid connection of large offshore wind farms *IEEE Int. Conf. on Power System Technology (POWERCON)* pp 1-6

[8] Nilanjan Ray Chaudhuri et.al 2014 *Multi-Terminal Direct Current Grids Modeling, Analysis and Control* (Willy)

[9] National Electric Power Regularity Authority 2016 *State of Industry Report* (Government of Pakistan)

[10] Shahzada Adnan et.al 2012 Solar Energy Potential in Pakistan *Journal of Renewable and Sustainable Energy* **4** pp 032701-032707

[11] Sajjad Haider Sham et.al Evaluating wind energy potential in Pakistan's three provinces, with proposal for integration into national power grid 2016 *Renewable and Sustainable Energy Reviews* **53** pp 408-421

[12] Umar K. Mirza et.al Status and outlook of solar energy use in Pakistan 2003 *Renewable and Sustainable Energy Reviews* **7** pp 501-514

[13] Samreen Siddique, Rashid Wazi 2016 A review of the wind power developments in Pakistan, *Renewable and Sustainable Energy Reviews* **57** pp 351-361

[14] National Transmission Dispatch Company (NTDC): www.ntdc.com.pk

[15] China Pakistan Economic Corridor (CPEC): www.cpec.gov.pk

[16] Kala Meah, Sadrul Ula 2007 Comparative Evaluation of HVDC and HVAC Transmission Systems *IEEE Power Engineering Society General Meeting* pp 1-7

IOP Conf. Series: Earth and Environmental Science **133** (2018) 012005 doi:10.1088/1755-1315/133/1/012005

Voltage droop Coordinating Control applied in UPFC and STATCOM system

Huang Junhui[1], Peng Zhuyi[1], Ni Chengjie[2], Xu Yiqing[2], and Xue Jiliang[2]
[1]State Grid Jiangsu Economic Research Institute
[2]99 Mid HeNan RD. HuangPu District, Shanghai, China

Abstract. When UPFC，unified power flow controller is applied with other FACTS into power grid, it is possible that the voltage controlled vibrates constantly to response to a sudden reactive power turbulent in grid if the parameters of these FACTS are not coordinating reasonably. Moreover, the reactive power generated by these equipment will intertwine unexpectedly. The article proposes a method named voltage-reactive power droop control to allow the reference voltage fluctuating around the rating voltage so that the vibration is reduced and the power distribution is improved. Finally, the article cite a electric-magnetic simulation by EMTDC models of east-China power grid to prove it effective when applied to improve the response characteristics to sudden turbulence in power grid.

1. Introduction

Modern power grid is developing to a target which is embodied with long distance transmission, high voltage class, and large transport capacity. Grid's sectional reactive power unbalance and unevenness of power flow is increasingly stressing. FACT (flexible alternative current transmission) is an effective means to solve the problem. UPFC(unified power flow controller) is a one of the most qualified and comprehensive equipment among the FACTs. It can monitor the reactive power and voltage and control them when necessary. Moreover, it can control the power plow of the transmission line which the UPFC is installed.

With the increasing number of the FACTs installed in the power grid, the connections and inter-influences between the FACTs are becoming increasingly obvious. Usually, the district requiring installing the FACTs is area where FACTs are installed densely and concentratedly. If the system operation did not regulate the coordination control between these FACTs and allowed the behaviors of them by control of themselves, detrimental consequences would be anticipated[2][3].

Passage[2] resolve the problem of mutual influence between reactive power control segment with other control parts through decoupled flow control. Passage[3] discusses the unreasonable allocation of parameters used to coordinating the controlling system between different FACTs. Finally, it proposes a strategy realizing the separated-level control. Passage[5] proposes a control smoothing the electric measures during connecting and disconnecting. Passage[6] proposes inner –circle current intersecting decoupled control strategy, which realizes the decoupled control of active power and reactive power. Passage[7] proposes a decoupled control strategy based on power flow control and prove the effectiveness of improving the power flow adjusting ability. Passage[10] propose a multi-purpose control strategy. A appended control devise is designed using fuzzy logic.

This passage proposes a voltage-reactive power droop control which allow voltage controlled purpose to move around rating value under different load curve using voltage-reactive power droop characteristic. The result of simulation by EMTDC proves that this strategy is effective to suppress the harmful phenomenon existing in responses to net grid disturbance when equipment parameters do not cooperate reasonably. It hinder the appearance of competing-reactive power and voltage fluctuation of controlled bus.

2. Fundamental principles and structure of UPFC based on MMC

2.1. Fundamental principles based on MMC-UPFC

UPFC connects a voltage source converter(VSC) to transmission lines through a shunt transformer and a voltage source inverter to transmission lines through series transformer. The converter and inverter interconnect each other through a back-to-back DC structure realizing the simultaneous control of voltage and power flow. The principle figure presents as fig 1.

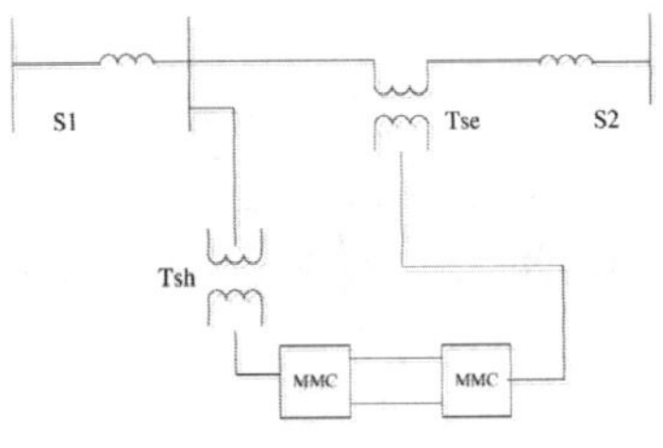

Figure 1. structure of UPFC

2.2. Topology structure of MMC-UPFC

Classical topology MMC structure is comprised of a series of power unites. Every phase consists two arms, the up arm and bottom arm. Each arm is comprised of the same number of power unites. The connection point of the two arms is the AC electric connection.

MMC's public DC bus does not need capacitors and other wave filters, which reduces the rate of fault and improves the reliability effectively. Moreover, MMC has large power capacity and fault ride through ability which ensures fast revise from fault condition and high efficiency of black-start.

UPFC based on MMC has stronger voltage and power flow control ability than traditional technology. Its structure is shown as follows.

MMC sub unite has two kinds of structures, full bridge and half bridge. This passage's simulation adopts half bridge structure. Half bridge structure consists of 2 capacitors controlled by IGBT. The structures of the electric circuits of half bridge is shown as fig 3.

IOP Conf. Series: Earth and Environmental Science **133** (2018) 012005 doi:10.1088/1755-1315/133/1/012005

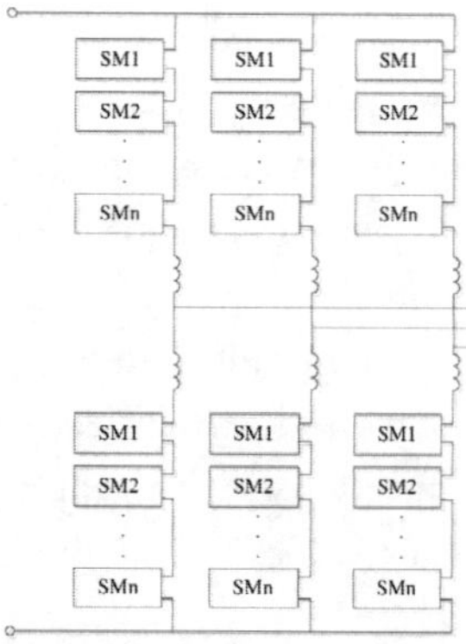

Figure 2. Diagram of UPFC MMC bridge structure

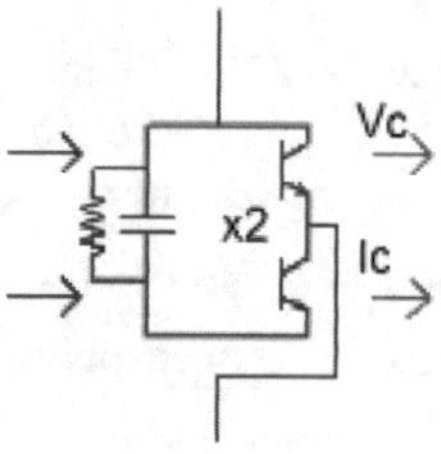

Figure 3. Sub-unit of UPFC MMC half bridge structure

3. UPFC control strategy

3.1. Control strategy of UPFC's parallel side

While UPFC exchanges both active power and reactive power through its parallel side, active power control is realized mainly through its series side. The outer loop control can adopt setting-DC voltage-control, setting-AC voltage-control or setting-reactive power-control. The diagram of control is shown as fig 4 and fig 5. To maintain the voltage stability during operation, setting-voltage-control mode is usually adopted. The inner loop control adopts current cross-decoupling control. In practical engineering environment, every control parameter is converted to that present in d-q coordinates in order to realize decoupling calculation. UPFC's series control diagram is shown as fig 6.

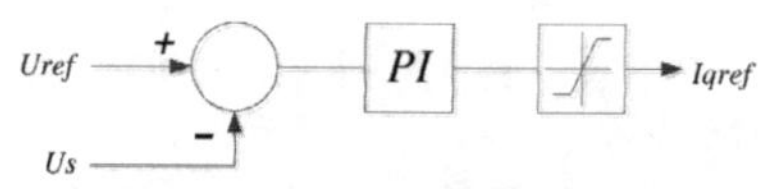

Figure 4. Diagram of constant AC voltage control

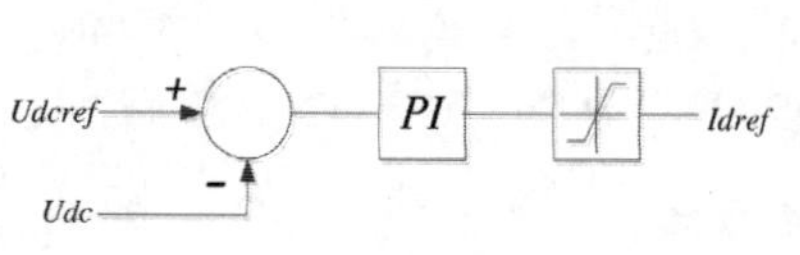

Figure 5. Diagram of constant DC voltage control

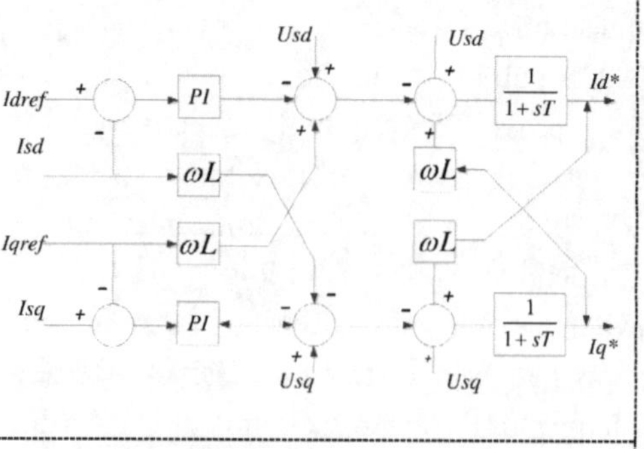

Figure 6. Diagram of inner current decoupling control

3.2. Control strategy of UPFC's parallel side

The purpose of control of parallel side is to accomplish power flow control and reactive power control. So, the outer loop is achieved by power loop empowered by PI section following the grid power signals. The outer control loop converts the power signals to current signals so that it is transferred to inner current control and accomplish close loop finally. The inner current loop is represented as fig 7.

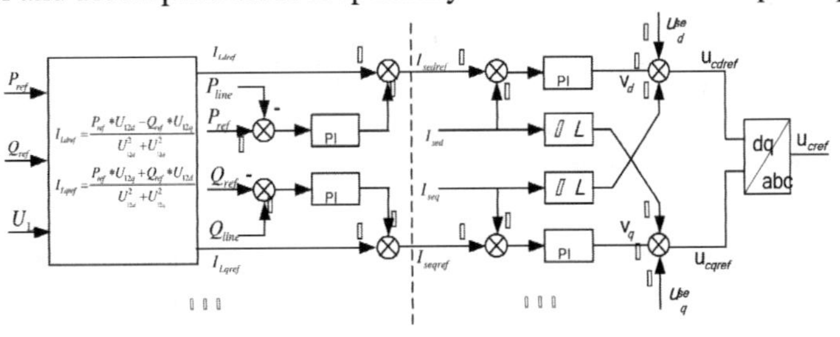

Figure 7. Diagram of inner current decoupling control in series sede of UPFC

4. Research of coordinating control of UPFC with other FACTs.

The model of system containing UPFC and statcom. The parallel side control of is of the same natural with stacom's. UPFC's voltage control of parallel side is not only related to valves of its own side but to the controlling purpose of series side. The active and reactive controlled at series side need to account for some amount of the parallel side's volume due to the transmission power flowing through DC section. The parameters of both sides will influence each other's corresponding speed[13].

When UPFC and statcom connect to the same connection point or when are placed close in a regional grid, they might exert controlling effect on the same or closely related electrical variables simultaneously. Omitting the leading, lagging segment and magnitude limit segment, the core part of statcom control is PI segment, whose main parameter is Kstatcom_p and Kstatcom_i. Considering the integral segment can be omitted when in the electromagnetic environment, the main parameter influencing corresponding speed and magnitude is Kstatcom_p. Similarly, the main parameter influencing UPFC's parallel control is the proportional parameter K_p, K_sp, K_sq. Consequently, parameter optimality is needed to coordinate well, operate independently and effect well when FACTs are placed closely.

Commonly, UPFC and statcom adopt voltage-setting control pattern to monitor grid voltage so that it is maintained within allowed range and provide timely reactive power support after the fault. However, if the main parameters like Kstatcom_p、K_p、K_sp、K_sq did not cooperate well, the effect described before might be magnified, which is presented as competing-reactive power and voltage fluctuation of controlled bus.

To solve the problem, two methods are proposed: 1, construct the problem as mathematically optimal problem with constrains and solve it through mathematical methods[10][20]. 2, adopt main-subordinate control in which one equipment adopts voltage-setting control(main) and the others adopt compensatory power-setting control(subordinate). Method 1 is complex and limited by specific grid, which lacks generation. Method 2 needs real time communication so it is limited by equipment reconstruct.

5. Voltage-droop control strategy

5.1. Summary of voltage-droop control strategy

Voltage-droop control strategy has universally been used in multi-terminal DC transmission system as the primary frequency modulation. To realize plug-and-use and PCC voltage stability in islanded pattern, voltage-droop control strategy is commonly adopted.

This passage proposes a voltage-reactive power droop control strategy applied in FACTs concentrated grid. Power control intends to control the amount of power generated by each equipment. Voltage-setting control intends to keep the voltage stable. Voltage-reactive power control combine these two types of characters and achieves the coordinating power generation of each equipment by allowing voltage fluctuating within a small range.

5.2. Control character and realization of voltage-droop control strategy

Voltage-droop control's character is presented as fig 8, $Q_{C\max}$ and $Q_{L\max}$ is the maximum and minimum infused reactive power in connection point. The Y-axis presents the voltage of the connection point. U_{r-set} is the controlling purpose when no reactive power is generated.

Control diagram of voltage-droop control is shown as fig 9. In the fig, the err exported is described as follows.

$$err = k_q(Q_{ref} - Q) + k_u(U_{ref} - U_s)$$

In the equation, k_q and k_u is the coefficient and the slope of the droop line is $-k_p / k_u$, when $k_q = 0$, droop control is reduced to voltage-setting control.

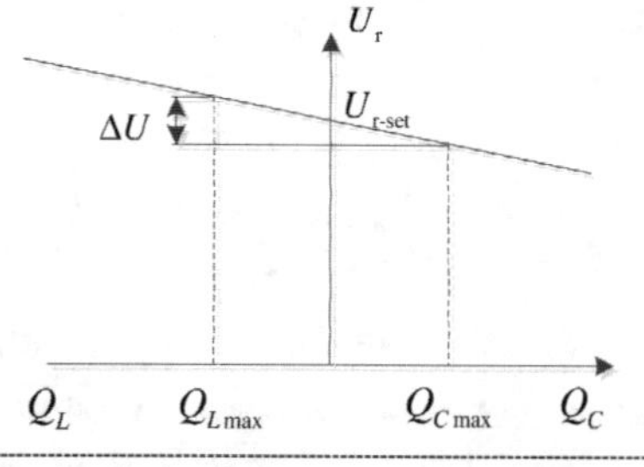

Figure 8. characteristic line of voltage-var droop control

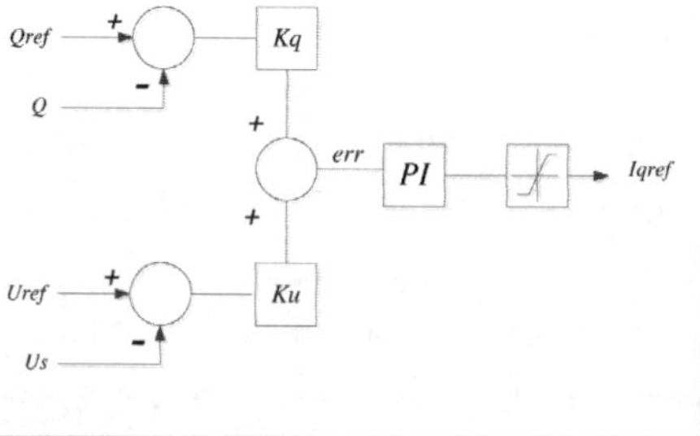

Figure 9. Control Diagram of droop control

6. Simulation and experiment consequences

This passage uses an electromagnetic model describing the power grid of southern Jiangsu province and shanghai city. The model not only includes these two districts but also include other part of the East China power grid, which is processed by dynamic equivalence. The whole grid includes 8 DC transmission systems and more than 1000 AC nodes. In this system, 3 reactive power generators are installed at node #531010(UPFC × 1, stacom × 2). UPFC's shunt rating capacity is 300MVA, statcom's capacity is 300Mvar. Before 7s, the three devices have been in stable operation. At 3s, the reactive load increases by 300MW and these devices begin to respond.

Situation1: none of the devices apply voltage-reactive power control, from fig 10, we can see that after 7s, bus voltage oscillates conspicuously. Fig 11 illustrate the appearance of competing-reactive power and voltage fluctuation of controlled bus.

Situation2: all the devices apply the voltage-reactive power control and the droop rate is -0.027, fig 10 shows that voltage operational point steps into new region and voltage oscillation is hindered. Fig 12 illustrates the average distribution of the free reactive generation devices and the hindered appearance of competing-reactive power and voltage fluctuation of controlled bus.

As is discussed above, voltage-droop control can hinder defects caused by parameter when equipment parameters do not cooperate reasonably.

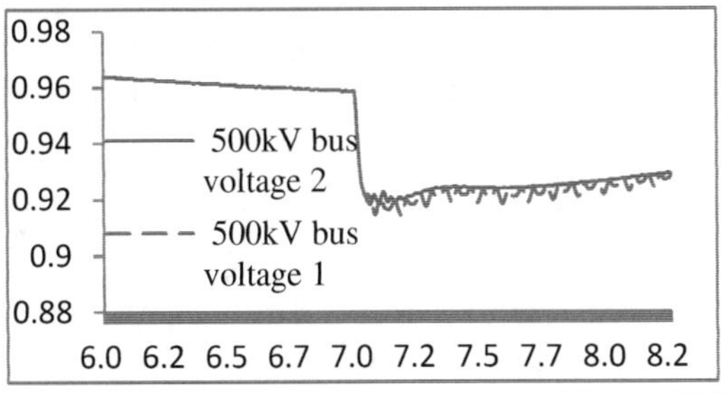

Figure 10. Voltage of controlled 500kV bus

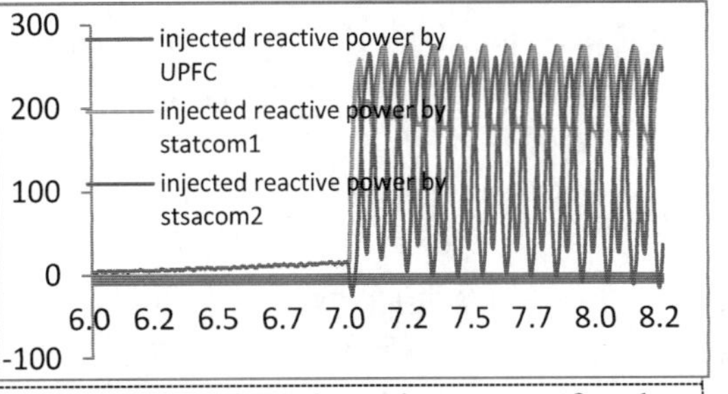

Figure 11. Injected reactive power of each equipment（without droop control）

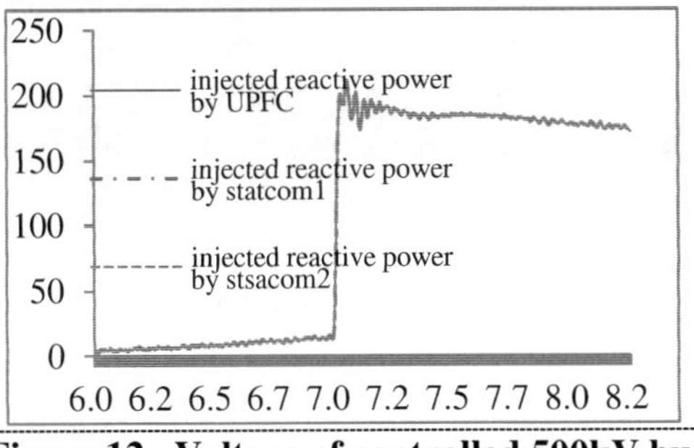

Figure 12. Voltage of controlled 500kV bus

References

[1] B Ren, W Du, H Wang, H Li, Z Chen. Dynamic interaction investigation on a UPFC connecting to Jiangsu UHV AC/DC hybrid power system[J]. Power System Technology[1] 2016, 40(9):2654-2660.

[2] LIU Qing, MA Peng, ZOU Jiaping, Solution to negative interaction caused by reactive power flow controlling of UPFC[J]. Power System Protection and Control, 2016, 44(6):76-81.

[3] XU Zhenhua1, LU Zhengang2, WU Danyue1, HUANG Daoshan1, Research on Coordinated Control Between Unified Power Flow Controller and Other FACTS Equipments[J]. Smart Grid2016, 4(1):25-32.

[4] WANG Weiyuan1，DOU Fei1，YANG Lin1，ZHOU Yuzhi2，XU Zheng2，CHEN Guonian1，YUAN Jianhua3, Applications and Control Results of UPFC in Jiangsu Power Grid[J], Power System and Clean Energy,2016(3):92-97.

[5] PANLei LIJihong TIANJie DONGYunlong SUN Weizhen ZHANGJing,SmoothStartandStopStrategiesforUnifiedPowerFlowControllers[J],Automation of Electric Power System. 2015, 39(12).

[6] CHEN Yong1, HUANG Shushu, Initial Allocation of Carbon Emission Right for East China[J], 2016, 35(1):45-48.

[7] LIU Li-ming, KANG Yong, CHEN Jian, ZHU Peng-cheng, Cross-coupling Control Scheme and Performance Analysis for Power Flow Control of UPFC[J], 2007, 27(10):42-48.

[8] DU Wen-Juan , WANG Hai-feng , M Jazaeri , JU Ping , CAO Yi-jia, Effect of Variations of Control Operating Points of UPFC on Power System Stability and Control Performance[J], Automation of Electric Power System. 2005, 29(20):40-45.

[9] XU Chen, DAI Ke, FANG Wuping, KANG Yong, Research on PCC Voltage Droop Control of STATCOM Based on Modular Multilevel Converters, Proceedings of the CSEE, 2015(s1):205-212.

[10] ZHANG Man1, ZHANG Chunpeng1, JIANG Qirong1, ZHOU Fei2, SONG Jieying2, Study on Multi-Objective Coordinated Control Strategy of Unified Power Flow Controller[J], Power System Technology,2014, 38(4):1008-1013.

[11] LIN Jinjiao, LI Peng, KONG Xiangping, GAO Lei, YUAN Yubo, HUANG Haosheng, WANG Ye The UPFC Protection System Configuration and Action Strategy[J],in Nanjing Western Power Gird, 2015, 34(6):56-60.

[12] L Huang，L Zhang，H Xin，J Hu，D Gan, Mechanism analysis of virtual power angle stability in droop-controlled inverters[J], 2016, 40(12):117-123

[13] LIU Qing, MA Peng, ZOU Jiaping, Solution to negative interaction caused by reactive power flow controlling of UPFC[J]. Power System Protection and Control, 2016, 44(6):76-81.

[14] ZHAOYuan,YANGXiaosong,XIEKaigui, parameter sensitivity and optimal allocation of UPFCs in bulk power systems reliability assessment[J]. Automation of electric power systems[J] 2012, 36(1):55-60.

[15] LI Haifeng, WANG Xiangfeng, CHEN Zhong, REN Bixing, DU Wenjuan, Quantitative Comparison Methods of a Power-Angle Controlled and Vector-Current Controlled UPFC[J], Power System Technology [J], 2016, 40(8):2330-2336.

[16] WANG Shuaibing, LI Lin1, XIE yuqing,ZHAO Guoliang,Research on SCT Characteristics and Simulation Model in UPFC System[J]. Power System Technology, 2016, 41(2):551-557.

[17] Noroozian M, Angquist L, Ghandhari M, et al. Use of UPFC for optimal power flow control[J]. IEEE Transactions on Power Delivery, 1997, 12(4):1629-1634.

[18] Guo J, Crow M L, Sarangapani J. An Improved UPFC Control for Oscillation Damping[J]. Power Systems IEEE Transactions on, 2009, 24(1):288-296.

[19] Tambey N, Kothari M L. Damping of power system oscillations with unified power flow controller (UPFC)[J]. Generation, Transmission and Distribution, IEE Proceedings-, 2003, 150(2):129-140.

[20] Kamarposhti M A, Lesani H. Effects of STATCOM, TCSC, SSSC and UPFC on static voltage stability[J]. Electrical Engineering, 2011, 93(1):33-42.

IOP Conf. Series: Earth and Environmental Science **133** (2018) 012006 doi:10.1088/1755-1315/133/1/012006

Design and implementation for integrated UAV multi-spectral inspection system

X Zhu, X Li, F Yan[+]
School of Electronic Science and Engineering, Nanjing University, Nanjing 210023, China

Abstract. In order to improve the working efficiency of the transmission line inspection and reduce the labour intensity of the inspectors, this paper presents an Unmanned Aerial Vehicle (UAV) inspection system architecture for the transmission line inspection. In this document, the light-duty design for different inspection equipment and processing terminals is completed. It presents the reference design for the information-processing terminal, supporting the inspection and interactive equipment accessing, and obtains all performance indicators of the inspection information processing through the tests. Practical application shows that the UAV inspection system supports access and management of different types of mainstream fault detection equipment, and can implement the independent diagnosis of the detected information to generate inspection reports in line with industry norms, which can meet the fast, timely, and efficient requirements for the power line inspection work.

1. Introduction

The stable operation of the transmission line is a prerequisite to ensure the complete operation and the social stability of the power system, which is a key point in the power distribution. As most of transmission line equipment is exposed to the complex environment of the outdoors over the long term, oxidization, broken strand, rust, overheating and other defective conditions may occur with the wires, lightning wires, insulators, fittings and other equipment. In order to ensure the normal operation of the transmission line, the inspector shall regularly inspect the fault of the transmission line. In order to avoid missed inspection, the inspector shall carry a variety of testing equipment for comprehensive testing [1]. Such comprehensive inspection work will consume a lot of manpower. Even with all kinds of testing equipment being carried, the instruments do not have the linkage characteristics to each other, and each instrument needs to detect each phenomenon one by one to find contrasts, which has low efficiency and is difficult to find potential problems. With the increase in the number of inspection equipment, the inspection workload is also increasing. Therefore, we urgently need a convenient, comprehensive and efficient inspection method.

UAV has advantages of easy carrying, simple operation, rapid response, abundant load, wide range of tasks, low environment requirements for take-off and landing, autonomous flight, etc. [2] The application of UAV to power inspections can greatly improve the speed and the efficiency of power maintenance and overhaul so that many jobs can be completed in a fully charged environment to ensure the electrical safety. The use of UAV in the conventional transmission line inspection can reduce the labor intensity, and can improve the safety of inspection operators and reduce the costs as compared with the inspection with a manned helicopter. UAV inspection is fast, has high speed in emergency, is capable of detecting defects, and provides information in a timely manner to avoid the power outages due to line accidents and restores the high losses caused by power outages. Through the joint line inspection with rotor wing and fixed wing UAVs, we can quickly inspect a wide range of transmission line corridor and efficiently have access to high-definition photos of the line equipment. This allows us to clearly distinguish the defects of the common transmission line, which greatly enhances the inspection efficiency of the transmission lines, and provides an advanced technical means with practical values for the maintenance of the modernized transmission line. For the infrared, UV detection and other tasks shot by UAV with visible light, the airborne equipment implements the flight inspection to the transmission line, and transmits the on-site situation back to the ground monitoring system in real time in order to make a correct judgment and promptly exclude line failures. It also shoots 360 degrees around the target and in all directions for zero dead angle.

In recent years, the means of optical detection have become more and more important in electric power detection systems. At present, the optical charged detection means used in the power system includes infrared imaging technology, UV imaging technology and visible light HD monitoring [3]. Because of different natures of the detected faults, these three detection techniques are often independent of each other in the hardware. In the power detection process, different devices are used to separately detect faults of the circuit system in different stages. Therefore, it is not conducive to the safe and efficient operation of the power detection. It is of practical significance to develop UAV-mounted UV/visible/thermal infrared composite system [4].

On the other hand, the UV detection technology has gradually developed into a newly-emerging photoelectric detection technology following the infrared and laser detection technologies. The core technology in UV detection -- solar-blind UV imaging technology is one of the most sophisticated military-purpose photoelectric detection technologies in the world. It uses "solar blind region" to reduce the background interference of ultraviolet radiation from the sun, and only needs to detect the UV radiation of the target, so as to enhance the speed of information processing. Ultraviolet light is the electromagnetic wave with the spectrum wavelength between 100nm and 400nm. Solar radiation background noise is natural noise. All ultraviolet detectors actually work in the context of solar spectral radiation. Figure 1 is a schematic diagram for solar-blind UV detector under the background of solar spectral radiation [5].

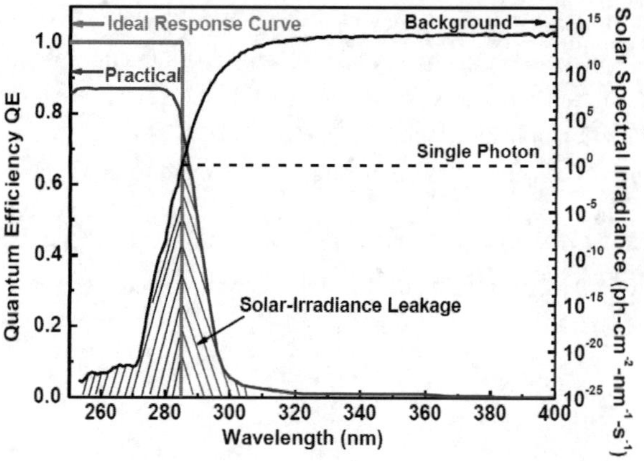

Figure 1. Schematic diagram for solar blind UV detector in the context of solar spectrum radiation [5]

The ideal detector response curve is sharply cut off at 285nm, and for the wavelength part afterwards, the background of solar radiation is negligible. But due to the decreased photon absorption rate near the cut-off wavelength, there is usually a drop tail in the detector response curve. And after wavelength exceeds 285nm, solar spectral radiation increases exponentially, and has reached $10^{13}ph/cm^2/nm^1/s^1$ at 300nm, so such a tail produces a large number of photons that cannot be ignored [6]. Even if we take into account the environmental reflection loss, the number of leaked photons is still very large. Therefore, solar radiation background noise is another major source of detector noise that cannot be ignored, in addition to the noise of the detector itself [7].
This article will design a light-duty solar-blind UV camera for the integrated UAV multi-spectral inspection system.

2. System design

As the UAV load has a certain limit, the inspection equipment shall be light-based, and the inspection analysis needs to have the automation capability. The design first considers maximizing the integration of inspection modes of all inspection equipment. It then considers the realization of interoperability of all types of inspection equipment, build a data information processing system, and considers the scalability of the entire system. Finally, it implements a lightweight design for each piece of inspection equipment.

2.1. Integration mode

This paper designs the integration mode applicable for UAV inspection, including UV, infrared and visible inspection equipment, based on the understanding of all types of equipment for power line inspection.

2.2. System architecture design

The whole architecture is divided into three parts, including the information collection equipment, the information-processing platform, and the remote terminal equipment. The information collection equipment, the information processing platform and the terminal are networked in the wired or wireless way, and the wired interface is a common communication interface, such as a network port, RS232/RS485, USB, etc., with strong universality. The information-processing platform contains the wireless network module, of which the function is used to communicate with the remote terminal to facilitate the processing of its equipment information collection, processing, management, storage and forwarding output services.

2.3. Data information processing system

After completing the integration mode of the inspection equipment, all the equipment needs to be interconnected, so that the whole system can play its value [8]. On the other hand, an information-processing platform is needed to collect, process, store, manage and output all the information of the testing equipment, thus reducing the function dependency of the system on the inspection equipment. There are many sets of detection equipment of a variety of data types in the transmission line inspection; the integrated information processing terminal should include multi-channel video, voice and data stream information processing, and has low power consumption, small size, strong performance, easy expansion and other properties.

IOP Conf. Series: Earth and Environmental Science **133** (2018) 012006 doi:10.1088/1755-1315/133/1/012006

2.3.1. Hardware platform selection

In all information types, multi-channel voice and data class information processing has relatively lower requirements for the platform, so the solution of the multi-channel video information processing is the key to the system. In the actual inspection, it should meet at least three-channel video information collection, algorithm operation, coding and network output, and need to have enough voice and universal data transmission interfaces for accessing of inspection equipment.

The information processing terminal usually uses the X86 or embedded platform. The single-chip embedded processor has been difficult to meet the performance requirements of this application, and when the ARM, DSP, FPGA and other integrated design are used, the difficulty to design the architecture and the development is surged. With the continuous development of the X86 hardware architecture, its power consumption and size have been gradually approaching to the embedded platform, and the X86 platform usually has a higher performance. In addition to the power consumption, size and performance, it is necessary to consider the development environment, openness and availability in the implementation. Many practical applications show that the X86 platform is superior in computing terminals [9].

In the specific selection, the Intel J1900 platform [10] is selected, and its dominant frequency is 2.0GHz, supporting Intel 7th-generation display core; in addition to the high dominant frequency, the GPU is used for parallel algorithm development and video hard coding. The entire board peak power consumption of Intel J1900 is 10W, and it has more common data peripherals to fully meet the requirements of the system.

2.3.2. Software architecture and design

In the integrated information terminal part, the software architecture diagram is shown in Figure 2.

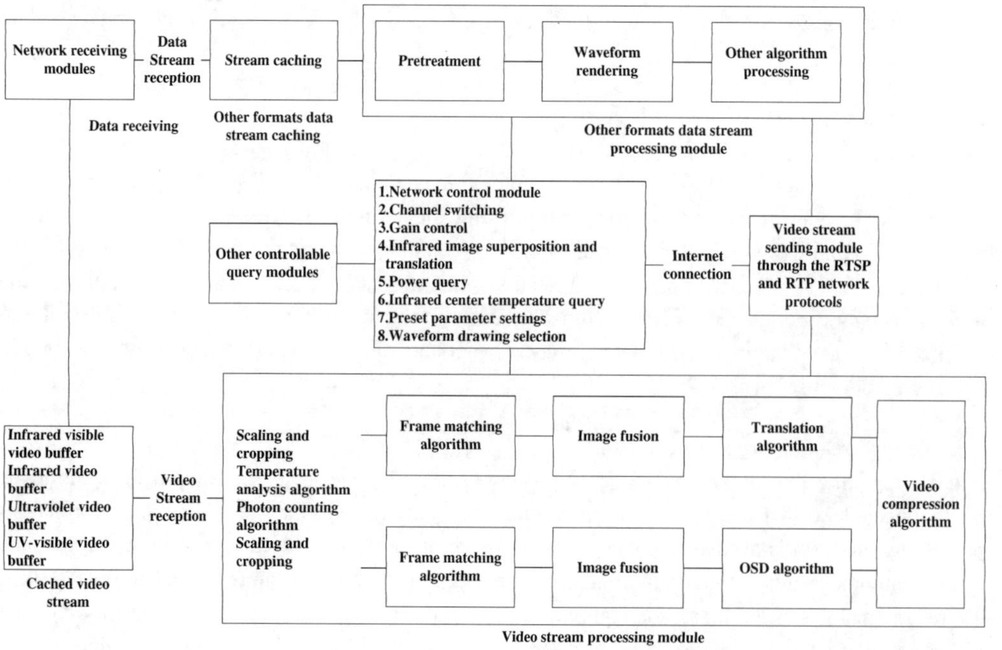

Figure 2. Software architecture diagram of the data information processing system

IOP Conf. Series: Earth and Environmental Science **133** (2018) 012006 doi:10.1088/1755-1315/133/1/012006

During the software architecture, it is divided into the information stream processing module and the control module. With the video stream, for example, the audio and data class multi-channel information processing and control modes completely refer to the software architecture of the video stream part. In the video stream processing module, first the multi-channel video stream of the module networking is buffered and received through the network, and then the data of each frame is extracted in the cache for processing. In the video stream control protocol, complete each video stream switching, parameter settings, and other operation based on the standard protocol in the built-in TCP server.

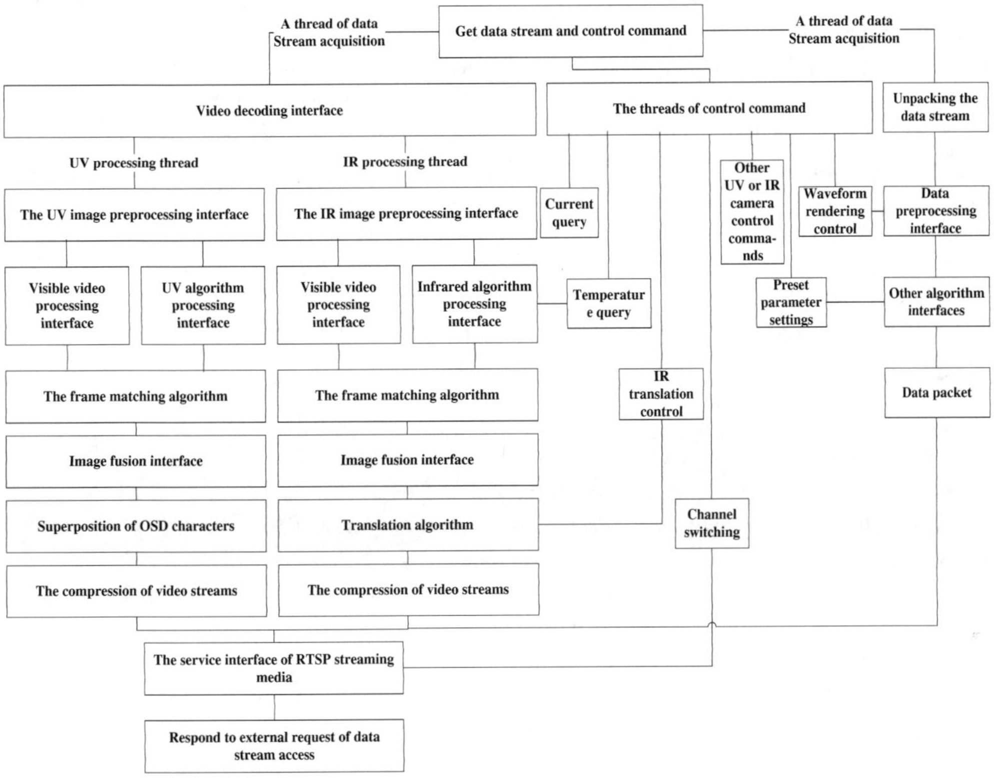

Figure 3. Software flow chart of the data information processing system

For each video stream, there are a customizable image preprocessing interface, an algorithm processing and analysis interface, an image fusion interface and a video compression interface. The image processing interface is mainly used for cutting, scaling, binarization and other processing. The algorithm processing and analysis interface is mainly used for algorithm analysis of different video streams, such as visible light video object recognition, infrared video temperature analysis, UV video photon counting and other algorithm. The image fusion interface is mainly used for information exchange and fusion between different devices. The video compression interface is mainly used for multi-channel H.264/H.265 compression to reduce the amount of information transmitted. In the video stream processing process, it can store the intermediate variables and the result files to achieve the

IOP Conf. Series: Earth and Environmental Science **133** (2018) 012006 doi:10.1088/1755-1315/133/1/012006

localized data storage, and also can be called by other devices through RTSP or RTP network protocol in real time. In addition, it provides the TCP server for internal running mechanism switching and parameter settings of other devices in real time.

During the specific development, the multi-threaded writing is used, the integrated development environment is VS2013, and the software flow chart is show in Figure 3.

After the software is initialized, the system opens a video stream acquisition thread for obtaining the video stream and the data stream of the access system. When the corresponding information is acquired, a service thread is created, the obtained video stream is decoded, the data stream is unpacked, and the processed information is pushed to the corresponding buffer channel. At this time, start the processing thread that has received the channel data for the UV, infrared and data processing.

While the video stream processing thread is running, the control command thread is running at the same time for real-time configuration and selection. The control thread is essentially a TCP server, and is used to monitor the control and query commands sent from the fixed port for parsing, verification, processing, and response.

3. Equipment light-duty design

At present, the infrared and visible UAV equipment has been basically matured, but the UV UAV equipment is still in the exploratory stage. This paper presents the UV detector miniaturization example and successfully implements based on the industry experience.

3.1. UAV airborne UV imaging module

The system chooses the optical material with the fused silica and calcium fluoride as the lens, and these two materials have good transmission performance to the solar blind UV band. After that, the UV lens was optically designed and optimized according to the requirements of UAV.

The UV module mainly consists of the UV CCD, UV lens, control circuit board, fixed structure and other sub-modules. Since the rigidity of the baseboard will directly affect the stability of the UV module, in order to increase the rigidity of the baseboard and reduce the assembly error, the baseboard of the UV module adopts the integrated design. All the sub-modules are directly fixed on the baseboard, so the assembly error is greatly reduced. Its strength is taken into account while reducing the weight. The baseboard is made of 7075 aviation aluminum material, and its wear resistance and corrosion resistance are increased through the surface oxidation.

At present, the overall weight of the UV module is less than 1kg, which can meet the needs of UAV. Figure 4 shows the physical diagram for the UV imaging module.

Figure 4. Physical diagram of the UV module

IOP Conf. Series: Earth and Environmental Science **133** (2018) 012006 doi:10.1088/1755-1315/133/1/012006

The technical parameters of the UV module are shown in Table 1, and the performance parameters have reached the international front.

Table 1. Technical parameter list of the UV module

Performance	Technical Parameter
Sensitivity	$2\times10^{-18}W/cm^2$
Working band	240-280nm
Perspective	10.4°× 7.8°
Focal length	78mm
Frame frequency/Hz	25
Weight	1kg
Focusing mode	AF mode

The test results of solar-blind performance and UV sensitivity are shown in Figure 5.

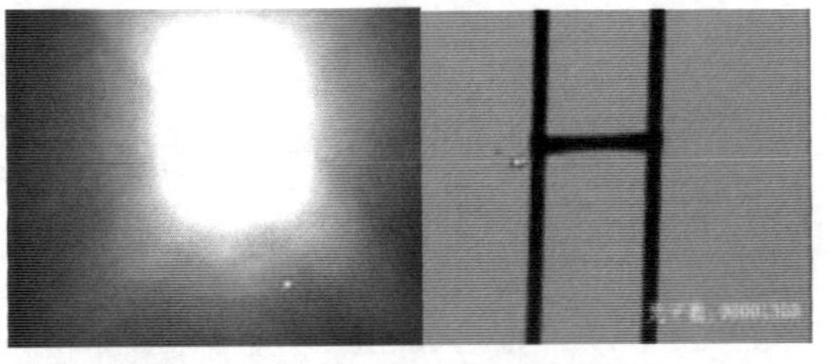

Figure 5. Solar-blind (left) and sensitivity (right) performance of UV module, which is working in combine mode (UV (white point) & Visible))

These results show that the performance of the UV module meets the requirements.

3.2. Infrared and visible module selection

At present, UAV infrared and visible imaging modules have been matured, and the infrared/visible module parameters used in the system are shown in Table 2 and Table 3.

Table 2. Technical parameter list of the infrared module

Performance	Technical Parameter
Sensor type	Non-cooled
Working band	8-14um
Perspective	8°×6°
Resolution	640×512
Frame frequency/Hz	50
Weight	200g
Focusing mode	AF mode

Table 3. Technical parameter list of the visible module

Performance	Technical Parameter
Focal length	Zooming
Resolution	1920×1080
Image size	1/4 in
Frame frequency/Hz	30

3.3. Information processing module

The information processing module mainly consists of the X86 platform, power management and network module. The information processing flow chart is shown in Figure 6.

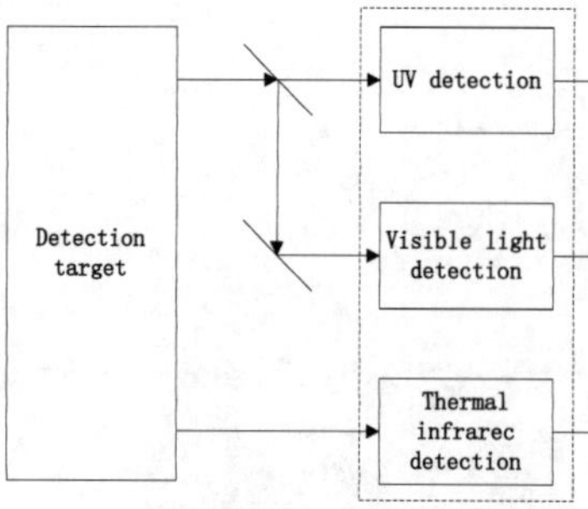

Figure 6. Information processing flow chart

The main functions in the processing terminal are divided into three parts: the data stream link, the control stream link and the processing container. Where the data stream link is mainly used for information collection, compression and transmission of the inspection equipment accessed to the system. The control stream link is mainly used for parameters setting, function switching and other interaction of the inspection equipment. The processing container is mainly used for the information processing, integration and analysis of the inspection equipment.

In all the information processed, the image and video information has the highest requirements for the platform. The related performance indicators for average 1080P image processing after several tests under the Intel J1900 platform are shown in Table 4.

Table 4. Image/video processing performance test table of the integrated information processing terminal

Item	Content	Time spent (fps)	Consumed memory
Video compression	H264	0.0297s	215.2Mb
	MPEG4	0.0343s	307.6Mb
Image processing	Crop	0.0173s	8.7Mb
	Shifting	0.0091s	9.2Mb
	Fusion	0.0315s	17.3Mb
Image analysis	Photon counting	0.0464s	8.5Mb
	Fault zone analysis	0.0382s	9.1Mb

4. Practical application

Figure 7 shows the physical diagram of the photoelectric pod, which is sealed with metal and window glass. Each module of the multi-spectral inspection system has a separate window.

Figure 7. Photoelectric pod

Figure 8 (left) shows the schematic diagram of the inspection site. Figure 8 (right) is a picture of the integrated UAV during inspection.

Figure 8. Schematic diagram of the inspection site (left) and integrated UAV during inspection (right)

Figure 9 shows sample of UV images. After the inspection, we find a burr where the UV image shows.

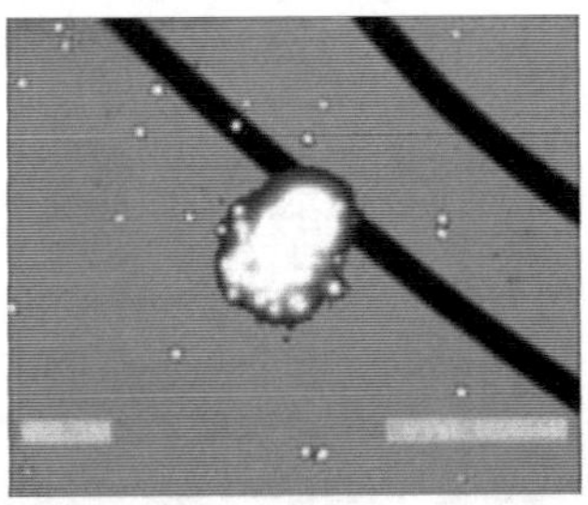

Figure 9. Sample of UV image

Figure 10 shows sample of VIS and IR images, where bad connection is found at the hot point shown in the IR image.

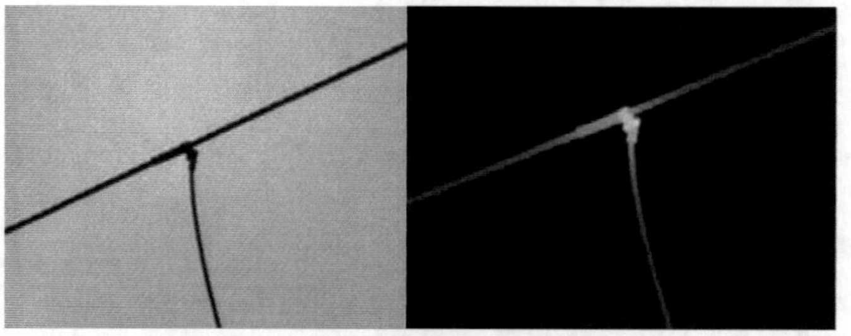

Figure 10. VIS (left) / IR (right) Image

When the inspection equipment is turned on, the handheld terminal has a connection notification and describes the relevant information of the equipment. On the handheld terminal, set the automatic fault analysis mode of the UV detector. When the questionable fault point is detected, the alarm information is displayed at the handheld terminal. It can meet the needs of all-day inspection. After the system is used, during the inspection, the inspection equipment is easy and fast to use, and the inspection human/times are significantly reduced; the inspection information on each equipment is under linkage response, and it can automatically identify and analyze the fault, and fault detection rate is greatly improved.

5. Conclusion and Outlook

This paper designed and implemented a set of UAV power inspection system, and presented a new and efficient power line inspection mode. A set of open hardware and software standard interface is designed for the entire inspection system, which can support access to the inspection equipment of different technical means for unified management, storage, analysis and report generation; the system can also support handheld terminals and remote network access.

This paper gives an example of UAV UV imaging equipment, and is more lightweight and internationally advanced while meeting the performance requirements of similar products in the market.

This paper gives the overall solution of the information processing platform in the inspection system, completes the platform selection, software design, development and debugging, and implements the full functions on the intel J1900 processing platform and tests out the specific performance indicators.

The UAV power inspection system realized in this paper greatly improves the efficiency and quality of the power line inspection work, satisfies the practical needs of the field operation and maintenance personnel and ensures the safe operation of the grid equipment.

References

[1] DL/T 741-2010 *Overhead Transmission Line Operating Procedure*

[2] Li L 2012 Discussion on UAV inspection technology on overhead lines *Changsha University of Science and Technology* p 44-47

[3] Xiong D 2014 Study and application of UAV inspection line for transmission lines *Wuhan University of Science and Technology* 2014

[4] Cao X, Yue T, Lin X, Lin S, Yuan X, Dai Q, Carin L, and Brady D 2016 Computational Snapshot Multispectral Cameras: Towards Dynamic Capture of the Spectral World *IEEE Signal Processing Magazine* 33 (5): 95-108, 2016

[5] Li X, Zhu C, Zhu X, Xu Z, Zhuang X, Ji X, and Yan F 2013 Background limited ultraviolet photodetectors of solar-blind ultraviolet detection *Appl.Phys.Lett.* 2013 103,171110

[6] Ding J, Li X, Zhu X, Cao X, and Yan F 2015 Solar-irradiated leakage of UV camera for daytime corona inspection *Electrical Insulation and Dielectric Phenomena (CEIDP), 2015 IEEE Conference on* 10.1109/CEIDP.2015.7352117

[7] Liang Y, Zhang S, Cao X, Lu Y, and Xu T 2017 Free-standing plasmonic metal-dielectric-metal bandpass filter with high transmission efficiency *Scientific Reports* 7 (4357), 2017

[8] Zhang B 2008 Design and implementation of intelligent inspection management system *Shenyang University of Technology*

[9] Xu Y 2010 Design and research of wearable computing terminal hardware based on X86 architecture *University of Electronic Science and Technology*

[10] 2013 How To Choose AMD and Intel CPU Processor *Computer and Network* 2013, 39(9)

Acknowledgments

I would like to express my gratitude to all those who have helped me during the writing of this thesis. I gratefully acknowledge the help of my supervisor Professor Feng Yan and Dr. Xiang Li. I do appreciate their patience, encouragement, and professional instructions during the development and application process of the integrated inspection system.

IOP Conf. Series: Earth and Environmental Science **133** (2018) 012007 doi:10.1088/1755-1315/133/1/012007

Study on Heat Transfer Agent Models of Transmission Line and Transformer

B Wang and P P Zhang*
School of Mechatronics Engineering, University of Electronic Science and Technology of China, Chengdu, SC 611731, China

Email:1016765655@qq.com, peipei.zhang@uestc.edu.cn

Abstract. When using heat transfer simulation to study the dynamic overload of transmission line and transformer, it needs to establish the mathematical expression of heat transfer. However, the formula is a nonlinear differential equation or equation set and it is not easy to get general solutions. Aiming at this problem, some different temperature change processes caused by different initial conditions are calculated by differential equation and equation set. New agent models are developed according to the characteristics of different temperature change processes. The results show that the agent models have high precision and can solve the problem that the original equation cannot be directly applied in some practical engineers.

1. Introduction

With economic development, the demand for electricity is continuous increasing. The rated capacities of existing transmission line and transformer do not meet the increase of demand for electricity. Meanwhile, the actual capacity of the existing power transmission and transformation equipment is often not fully utilized when it is in normal operation, and there are still hidden capacity available, so it has become the best way to meet the power demand that the equipment is scientifically and safely dynamic overload [1-3].

Temperature of transmission line and insulation temperature inside the transformer are the key factors that directly affect their load capacity. At present, direct temperature measurement and indirect temperature measurement can get the transmission line temperature and transformer hot spot temperature. In the direct temperature measurement, sensors were directly installed on the transmission line and the transformer and monitored their temperature in real time [4, 5]. Besides, using the relationship between the wire temperature and its impedance, the temperature was calculated by directly measuring the line current and voltage [6, 7]. Indirect temperature measurement mainly includes heat transfer simulation [4, 8], numerical calculation [9, 10] and intelligent calculation [11, 12] methods. However, this mathematical expression of heat transfer simulation is a nonlinear differential equation or equation set, and it is not easy to obtain its general solution directly. Based on the original theoretical model, the new heat transfer agent models of transmission line and transformer are derived in this paper. The change process of the new models object approximately equal to that of the original models under the conditions that the initial conditions of both original models and agent models are the same, which solves the problem that the original models cannot be directly applied in some practical engineers.

2. Heat transfer characteristics of transmission line

2.1. Heat transfer transient differential equation of transmission line

The maximum allowable temperature of a transmission line is an important factor which limits the power carried by the transmission line. The temperature depends on various factors, including the current level, sunshine intensity, wind speed and ambient temperature. For heat transfer of transmission line, firstly solar radiation and current flows through the resistor make the transmission line generate thermal energy. Secondly some of the thermal energy is stored inside the transmission line, causing it to rise in temperature. The other part is dissipated by the convection of the wind and its radiation to the surrounding environment.

Transmission line current and the external environment are constantly changing, the heat transfer transient differential equation can be expressed as follows in equation (1) [8].

$$\frac{dT_e}{dt}=\frac{P+P_{st}-P_{rad}-P_{conv}}{MC} \tag{1}$$

Where $P=I^2\cdot R_T$, $P_{st}=\gamma\cdot D\cdot S_I$, $P_{conv}=9.92\cdot(T_e-T_0)\cdot(V\cdot D)^{0.485}$ and $P_{rad}=s\cdot\pi\cdot D\cdot k_e\cdot(T_e^4-T_0^4)$ respectively, and M is the mass per unit length of the transmission line [kg/m]; C is the specific heat capacity of the wire [J/(kg·°C)]; P is the power of thermal energy production due to the current per unit length of the transmission line [W/m]; P_{st} is the power of sunshine heat absorption per unit length of the transmission line [W/m]. P_{conv} is the power of convection cooling per unit length of the transmission line [W/m]; P_{rad} is the power of heat dissipation to the surroundings per unit length of the transmission line [W/m]; I is the current carried by the transmission line [A]; R_T is the AC resistance per unit length of the transmission line [Ω]; γ is the surface heat absorption coefficient; D is the diameter of the line [m]; S_I is the sunshine intensity [W/m^2]; T_e is the temperature of the transmission line [K]; t is the time [s]; T_0 is the ambient temperature [K]; V is the wind speed [m/s]; s is the Stephen-Boltzmann constant (5.67×10^{-8}); k_e is the radiation coefficient, and the bright new line is 0.23 ~ 0.46, while the dark degenerated line is 0.9 ~ 0.95.

2.2. Agent model of heat transfer in transmission line

Equation (1) is a first-order nonlinear differential equation, not easy to find its general solution, but a particular transmission line can be used numerical solution given approximate solution. For ACSR LGJ400/35 (steel wire wrapped in aluminum wire), the aluminum cross sectional area is 400mm^2, and the steel cross sectional area is 35mm^2. The M is 1.349kg/m; The C is 880J/(kg·°C); The R_T is 0.00007389Ω; The T_0 is 40℃; The V is 0.5m/s; the S_I is 1000W/m^2; The k_e is 0.9; The γ is 0.9; The outer diameter is 0.02682m; The line voltage is 138kV. Several different initial conditions are given, which are different initial temperature and load power. Their temperature change process is shown in figure 1.

IOP Conf. Series: Earth and Environmental Science **133** (2018) 012007 doi:10.1088/1755-1315/133/1/012007

In figure 1, the first "*" from the left of each curve is the node at which it changes from the initial temperature to a quarter of the difference from the initial temperature to steady-state temperature. The abscissas of first "*" of each curve are almost the same. The second "*" from the left of each curve is the node at which it changes from the initial temperature to a half of the difference from the initial temperature to steady-state temperature. The abscissas of second "*" of each curve are also almost the same. The n-th "*" from the left of each curve is the node at which it changes from the initial temperature to 1-1/(2n-1) of the difference from the initial temperature to steady-state temperature. Meanwhile, the abscissas of n-th "*" of each curve are almost the same (figure 1). According to this feature, a linear piecewise function can be constructed as its agent model. The coordinates of the segment nodes are shown in table 1.

Table 1. Point of subsection of approximate function of heat transfer of transmission line

	node1	node2	node3	node4	node5	node6	node7
X-axis	125	375	750	1250	1625	2000	2775
Y-axis	T_p+(1-0.25)ΔT	T_p+(1-0.5)ΔT	T_p+(1-0.5^2)ΔT	T_p+(1-0.5^3)ΔT	T_p+(1-0.5^4)ΔT	T_p+(1-0.5^5)ΔT	T_p+(1-0.5^6)ΔT

Where, $\Delta T = T_s - T_p$, T_s is the steady-state temperature, T_p is the initial temperature. The equation (1) is assumed to equal to 0, and the steady-state temperature can be obtained. The steady-state temperature is calculated per 1MW at load power from 1MW to 200MW. According to the obtained scattered values, the function have fitted with the binomial function, which is as shown in equation (2).

$$T_s = a \cdot P_1^2 + b \cdot P_1 + c \tag{2}$$

Where, a is 0.0006, b is 0.0043, c is 325.4.

In this way, it is possible to construct an agent model as shown in equation (3), and at a given initial temperature and the load power, an approximate function of the temperature at any time can be obtained.

$$T_s = f(t, T_p, P_1) = \begin{cases} T_p + [0.25 \cdot (T_s - T_p)]\dfrac{t-0}{t_1-0}, \ t \in [0, t_1) \\ [T_p + 0.25 \cdot (T_s - T_p)] + [(0.5^1 - 0.25) \cdot (T_s - T_p)]\dfrac{t-t_1}{t_2-t_1}, \ t \in [t_1, t_2) \\ \ldots\ldots \\ [T_p + (1-0.5^5) \cdot (T_s - T_p)] + [(0.5^5 - 0.5^6) \cdot (T_s - T_p)]\dfrac{t-t_6}{t_7-t_6}, \ t \in [t_6, t_7) \\ T_s, \ t \in [t_7, +\infty) \end{cases} \tag{3}$$

3. Heat transfer characteristics of a transformer

3.1. Heat transfer transient differential equations of transformer

The internal heat transfer process of the transformer is not same with transmission line, and the temperature of the iron cores and windings, the transformer oil and the housing are quite different, as shown in figure 2.

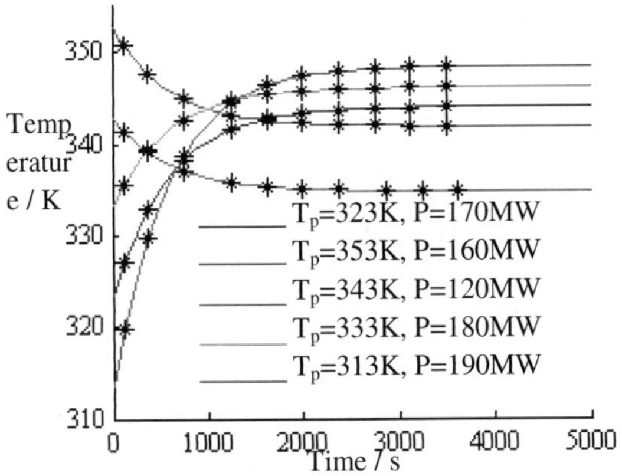

Figure 1. Temperature variation process of LGJ400/35 transmission line under several different initial conditions.

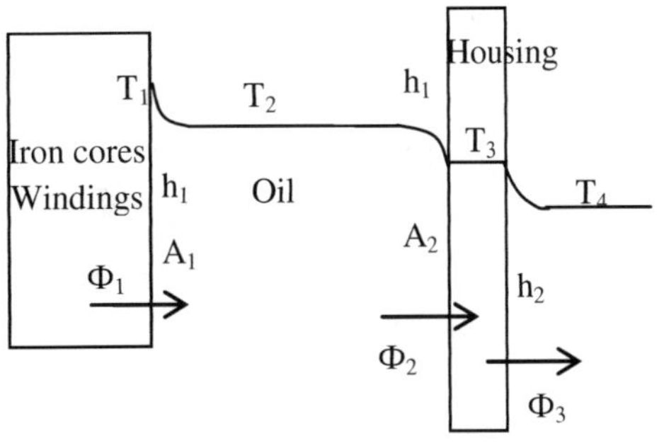

Figure 2. Heat transfer of a transformer sketch map.

First, the iron cores and windings generate thermal energy due to power loss. Some of the heat, Φ_1, is transferred to the oil due to the convection heat transfer. Some of the heat in the oil, Φ_2, is dissipated by the convection heat transfer to the transformer housing. The heat in the housing, Φ_3, is delivered to the surrounding air due to convection and thermal radiation. From the above heat transfer process, the heat transfer differential equations of transformer can be written as showing in equation set (4) - (6).

$$M_{\mathrm{I}} \cdot C_{\mathrm{I}} \cdot \frac{dT_1}{dt} = P - A_1 \cdot h_1 \cdot (T_1 - T_2) \tag{4}$$

$$M_{\mathrm{o}} \cdot C_{\mathrm{o}} \cdot \frac{dT_2}{dt} = [A_1 \cdot h_1 \cdot (T_1 - T_2) - A_2 \cdot h_1 \cdot (T_2 - T_3)] \tag{5}$$

$$M_{\mathrm{s}} \cdot C_{\mathrm{s}} \cdot \frac{dT_3}{dt} = [A_2 \cdot h_1 \cdot (T_2 - T_3) - A_2 \cdot h_2 \cdot (T_3 - T_4) - s \cdot A_2 \cdot k_e \cdot (T_3^4 - T_4^4)] \tag{6}$$

Where C_{I} is the specific heat capacity of the iron cores and windings [J/(kg·°C)]; C_{o} is the specific heat capacity of the oil [J/(kg· °C)]; C_{s} is the specific heat capacity of the housing [J/(kg· °C)], M_{I} the mass of the iron cores and windings [kg]; M_{o} is the mass of the oil [kg]; M_{s} is the mass of the housing [kg]; A_1 is the effective heat dissipation area of the iron cores and windings [m^2]; A_2 is the effective heat dissipation area of the housing [m^2]; h_1 is the convection heat transfer coefficient of the iron cores and windings [W/(m^2·°C)]; h_2 is the convection heat transfer coefficient of the housing [W/(m^2·°C)]; P is the power loss [W]; T_1 is the temperature of the iron cores and windings [K]; T_2 is the temperature of the oil [K]; T_3 is the temperature of the housing [K]; T_4 is the ambient temperature [K].

3.2. Agent model of heat transfer in the transformer

Equation set (4) - (6) are first-order nonlinear differential equation set. It is more difficult to solve than equation (1), but it can also be solved by numerical solution under the given initial conditions. For

S11-M-type transformer, its iron core is silicon steel sheet; the windings are oxygen-free copper; the windings interlayer insulating material is rhombus adhesive tape; the rated capacity is 400kVA; The maximum iron cores and windings temperature is 15°C higher than the average temperature of the transformer; The maximum temperature should not exceed 95°C. To ensure long-term operation, the average temperature should not exceed 80°C. The M_I is 295kg; The C_I is 450J/(kg·°C); The P is 7500(0.0125×600×10^3)kW; The A_I is 0.7m^2; The h_I is 500 W/(m^2·°C); The M_o is 280 kg; The C_o is 2060 J/(kg·°C); The A is 5m^2; The h_1 is 500 W/(m^2·°C); The M_s is 600kg; The C_s is 400J/(kg·°C); The h_2 is 50W/(m^2·°C).

Under five different initial conditions, their temperature changing curves are shown in figure 3.

Similar to the transmission line, an agent model is developed as shown in equation (3). The coordinates of the segment nodes are shown in table 2.

Table 2. Point of subsection of approximate function of heat transfer of transformer

	node1	node2	node3	node4	node5	node6	node7
X-axis	300	1225	3650	6131	8568	10970	13410
Y-axis	T_p+(1-0.25)ΔT	T_p+(1-0.5)ΔT	T_p+(1-0.5^2)ΔT	T_p+(1-0.5^3)ΔT	T_p+(1-0.5^4)ΔT	T_p+(1-0.5^5)ΔT	T_p+(1-0.5^6)ΔT

The equation set (4) - (6) are assumed to equal to 0, the steady-state temperature of the transformer can be obtained. The steady-state temperature is calculated per 1kW at load power from 1kW to 600kW. The steady-state temperature function (T_s) is fitted with a linear function according to the obtained scattered values, which is as shown in equation (7).

$$T_s = 0.08433P + 313 \tag{7}$$

In this way, the linear piecewise function can be constructed as an agent model of the temperature change process of the iron cores and windings. Given the initial temperature of iron cores and windings and the load power of the transformer, the temperature at any time can be calculated by equation (3) , Where t_1, t_2, ..., t_7 nodes are shown in table 2, T_s is as shown in equation (7).

For the S11-M transformer in this paper, in order to verify the accuracy of the agent model (3), two different initial conditions are selected to draw the theoretical temperature changing curve and the approximate temperature changing curve as shown in figure 4.

For the two kind of initial conditions (the 1st one and the 2nd one in figure 4), a series of time nodes are selected. The theoretical and approximate temperature values for the 1st one are compared as shown in table 3. From figure 4 and table 3, it can be seen that the errors of agent model at different time in table 3 are less than 0.1%, indicating that the agent model has higher accuracy. The error is calculated by '(proximate temperature-theoretical temperature)/theoretical temperature'.

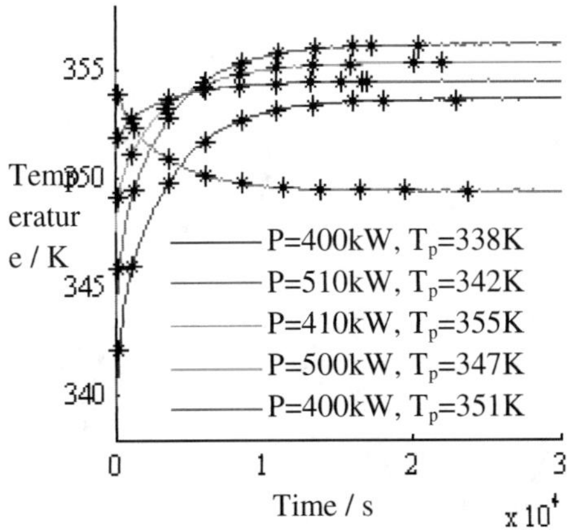

Figure 3. Temperature variation process of iron cores and windings of S11-M transformer under several different initial conditions.

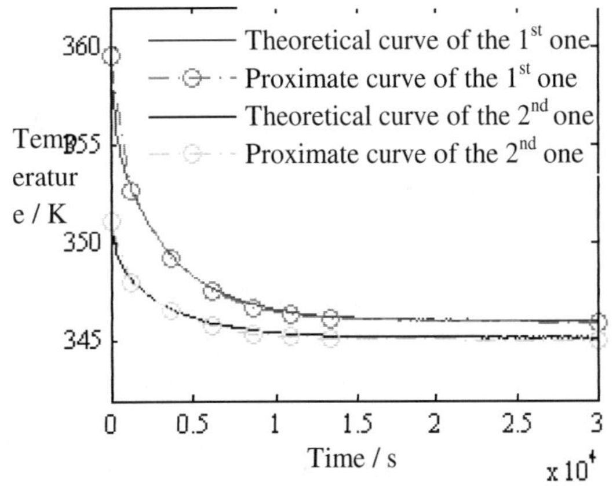

Figure 4. Theoretical temperature variation curve and proximate temperature variation of iron cores and windings of S11-M transformer.

Table 3. Theoretical temperature and approximate temperature of iron cores and windings for the 1st one

Time (s)	500	1000	2000	3000	5000	10000	15000
Theoretical temperature (K)	347.071	348.812	350.730	352.051	353.827	355.577	356.002
Proximate temperature (K)	346.747	348.546	350.418	351.791	353.588	355.426	355.822
Relative temperature (K)	-0.325	-0.266	-0.312	-0.260	-0.240	-0.151	-0.180
Error	-0.094%	-0.076%	-0.089%	-0.074%	-0.068%	-0.042%	-0.051%

4. Conclusions

In this paper, explicit agent function models of heat transfer of transmission line and transformer have been established respectively. After an initial temperature and load power are given, the temperature at any time can be calculated. The results show that the error between the agent model and the theoretical model is very small (in table 3 and figure 4), which can be applied to dynamically confirm capacity.

Acknowledgments

The work described in this paper was supported by key project of the National Natural Science Foundation (51537010).

References

[1] Song S X 2016 *Study of Dynamic-overload system based on hot spot temperature in oil-immersed transformer* (Jinan: Shandong University) pp 4-10 (in Chinese)

[2] Coletta G, Vaccaro A and Villacci D 2017 A review of the enabling methodologies for PMUs-based dynamic thermal rating of power transmission lines *Electric Power Systems Research* **152** pp 257-70

[3] Zhu L, Li H, Yao X and Ji S 2014 Improvement and application of dynamic thermal circuit model for oil-immersed power transformer. *Int. Power Modulator and High Voltage Conf. (Santa Fe)* (New York: IEEE) pp 591-4

IOP Conf. Series: Earth and Environmental Science **133** (2018) 012007 doi:10.1088/1755-1315/133/1/012007

[4] Kosec G, Maksić M and Djurica V 2017 Dynamic thermal rating of power lines – model and measurements in rainy conditions *International Journal of Electrical Power & Energy Systems*. **91** pp 222-9

[5] Seppa T O 2002 Increasing transmission capacity by real time monitoring *Power Engineering Society Winter Meeting (New York)* (New York: IEEE) pp 1208-11

[6] Du Y and Liao Y 2011 Online estimation of power transmission line parameters, temperature and sag *North American Power Symp. (Boston)* (New York: IEEE) pp 1-6

[7] Gubeljak N, Banić B, Lovrenčić V, Kovač M and Nikolovski S 2016 Preventing Transmission Line damage caused by ice with smart on-line conductor Monitoring *Int. Conf. on Smart Systems and Technologies (Osijek)* (New York: IEEE) pp 155-63

[8] Wang H B, Chen Y, Gao Y, Zhang J J, Geng D Q and Wen F S 2014 Impacts of transmission line sag on dynamic capacity increase *Journal of North China Electric Power University*. **41** pp 41-6 (in Chinese)

[9] Askari M T, Kadir M Z A A and Jasni J 2012 A numerical method for calculating the loss of life of power transformers *Int. Power Engineering and Optimization Conf. (Melaka)* (New York: IEEE) pp 29-34

[10] Lecuna R, Delgado F, Ortiz A, Castro P B, Fernandez I and Renedo C J 2015 Thermal-fluid characterization of alternative liquids of power transformers: A numerical approach *IEEE Transactions on Dielectrics and Electrical Insulation* **22** pp 2522-9

[11] Pradhan M K and Ramu T S 2004 Online monitoring of temperature in power transformers using optimal linear combination of ANNs *Int. Symp. on Electrical Insulation (Indianapolis)* (New York: IEEE) pp 70-3

[12] Cui Y, Ma H and Saha T 2015 Transformer hot spot temperature prediction using a hybrid algorithm of support vector regression and information granulation *PES Asia-Pacific Power and Energy Engineering Conf. (Beijing)* (New York: IEEE) pp 1-5

IOP Conf. Series: Earth and Environmental Science **133** (2018) 012008 doi:10.1088/1755-1315/133/1/012008

Energy Saving Performance Analysis of An Inverter-based Regenerative Power Re-utilization Device for Urban Rail Transit

Jin LI, Zhiling QIU and Leilei HU
Nanjing APAITEK Technology Co. Ltd, Nanjing, China, 210000

Abstract. The inverter-based regenerative braking power utilization devices can re-utilize the regenerative energy, thus reduce the energy consumption of urban rail transit. In this paper the power absorption principle of the inverter-based device is introduced, then the key influencing factors of energy saving performance are analyzed based on the absorption model. The field operation data verified that the control DC voltage plays an important role and lower control DC voltage yields more energy saving. Also, the one year energy saving performance data of an inverter-based re-utilization device located in NanJing S8 line is provided, and more than 1.2 million kWh energy is recovered in the one year operation.

1. Introduction

Recently, the urban rail transit gains rapid development due to its high capacity, punctuality and absence of emission. However, the urban rail transit consumes huge amount of power in its operation, and the energy saving issue has become an important topic for the sustainable development of urban rail transit.

During the regenerative braking of the urban rail transit train, the kinetic energy is converted into regenerative energy and fed back to the DC traction network. The regenerative energy, which accounts for about 30%-50% of the total traction energy consumption and offers the greatest energy saving potential[1-2], could boost the DC traction network voltage and may cause the regenerative braking failure so that it is required to be absorbed or utilized[3].

There are two state-of-art regenerative energy utilization devices, one is the storage-based device which stores the regenerative energy in the energy storage medium like super capacitors when the train is braking, and release the energy when the train is accelerating[4-5]. The other one is the inverter-based utilization device which can feedback the regenerative energy into the AC medium-voltage ring network and provide the power to the other metro line loads like traction system, environment control system, and etc. The inverter-based device becomes popular in China due to its features of higher power, smaller occupation and better performance cost ratio compared with storage-based utilization device[6-7]. In this paper, the inverter-based device is analyzed in terms of working principle, power absorption model, energy-saving performance.

2. Working principle and modeling

2.1. The principle of regenerative power utilization

As shown in the Figure 1, the inverter-based device is interfaced with the DC traction network and the medium voltage AC grid, and is parallel with the traction rectifier device. When the train is braking, a part of the regenerative power can be absorbed by the adjacent accelerating trains, and the rest of the power is fed back to the AC grid by the inverter and utilized by other loads.

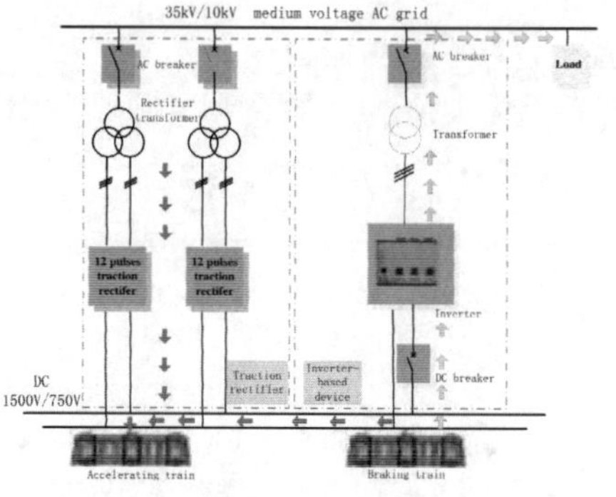

Figure 1. The system structure of substation with inverter-based utilization device

As shown in Figure 2 the control objectives of the inverter-based utilization device are to control the overhead line DC voltage and regulate the current injected into the AC grid. Seen from the DC traction network the inverter-based device behaves as a DC voltage source, while seen from the AC grid, the device behaves as a AC current source.

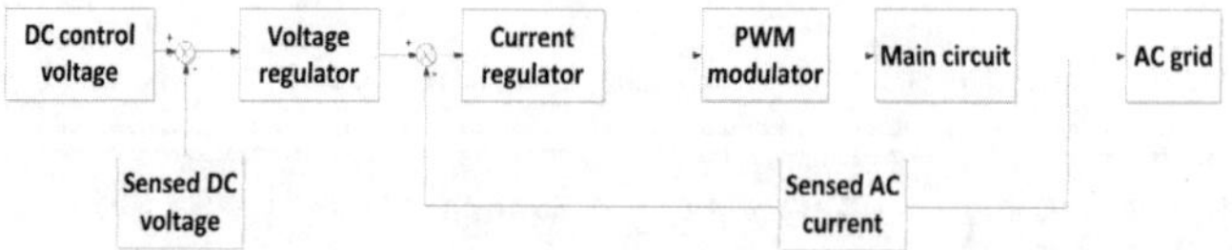

Figure 2. the control strategy of inverter-based device

In some cases, the on-board resistor-based devices as well as the inverter-based devices are equipped in the metro line. Since the DC control threshold voltage of the inverter-based device $V_{th\_inverter}$ is set be lower than that of the resistor-based device $V_{th\_resistor}$, the inverter-based device has the priority to absorb the power. Figure 3 shows the typical power absorption process waveform, at t0 instant, the inverter-based device starts to absorb the power. During the t0-t1 interval, the regenerative power is within the rated power of the inverter-based device, so the device can absorb all the regenerative energy and control the DC line voltage independently. at t1 instant, the regenerative power exceed the rated power of the inverter-based device, then the inverter-based device keeps outputting its rated power, and the exceeding power boosts the the DC overhead line voltage and reaches $V_{th\_resistor}$ and the

IOP Conf. Series: Earth and Environmental Science **133** (2018) 012008 doi:10.1088/1755-1315/133/1/012008

resistor-based device start to absorb the excessive power and stabilize the overhead line DC voltage. During t2-t3 interval, the regenerative power drops within the rated power of the inverter-based device, then the resistor-based device quits and all of the power is absorbed by the inverter-based device.

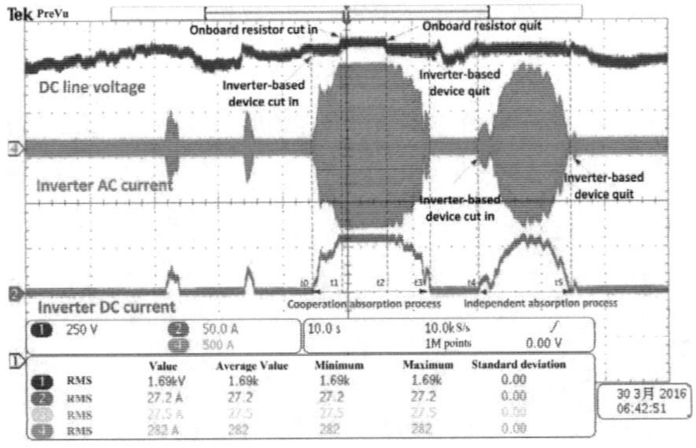

Figure 3. The typical waveform of power absorption process

2.2. Modeling of DC line regenerative power absorption model

As stated in the last section, the inverter-based utilization device is controlled as a DC voltage source, and the braking train can be modeled as a current injection into the overhead DC line. Since there are resistance between the braking train and the inverter-based utilization device located substation, the injection current could incur a voltage drop across the overhead DC line and rail track.

Figure 4 shows the flowing path of the regenerative current I_{cl}. As shown, the control DC voltage of the inverter-based device located substation is U_{dc}, and the DC voltage of the braking train U_{cl} can be expressed in equation (1), where μ_a and μ_b are the resistance rate of the overhead line and the rail track respectively where μ_a is 0.007Ω/km, μ_b is 0.009Ω/km, and d is the the distance between the braking train and the substation.

$$U_{cl} = U_{dc} + I_{cl} \cdot (\mu_a + \mu_b) \cdot d \tag{1}$$

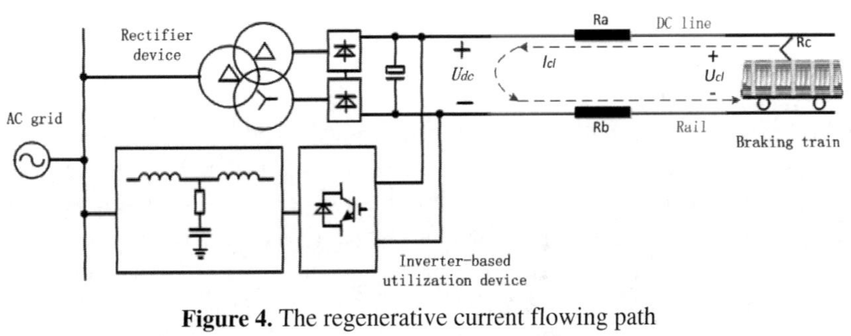

Figure 4. The regenerative current flowing path

So the inverter-based device regenerative absorption power is

$$P_{cl} = U_{dc} \cdot I_{cl} \tag{2}$$

It is assumed that the maximum allowable DC voltage for the regenerative braking train is 1800V, and the maximum absorption power of inverter-based device can be derived as equation(3).

$$P_{cl\_max} = \frac{1800 - U_{dc}}{(\mu_a + \mu_b)\cdot d} \cdot U_{dc} \tag{3}$$

There are two key factors influencing the maximum absorption regenerative power, one is the control DC voltage of the inverter-based utilization device, the other one is the distance between the braking train and the substation.

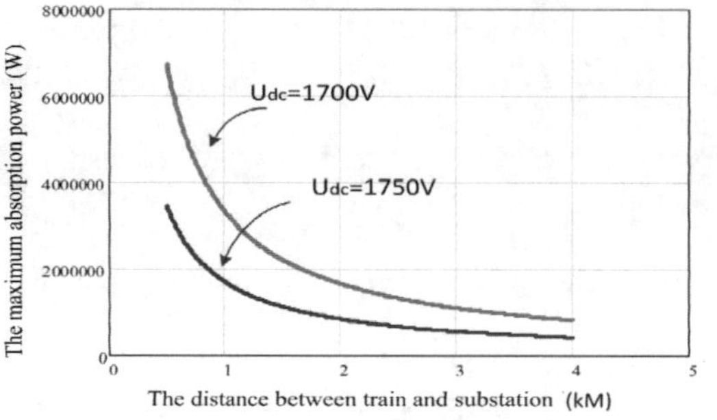

Figure 5. The maximum absorption power at different control DC voltage

Figure 5 shows the relationship of the maximum absorption power and the distance between the braking train and the substation with different control DC voltages of 1750V and 1700V. The lower control DC voltage results in higher absorption power, the maximum absorption power is about 6.7MW at 0.5kM braking distance with the control DC voltage of 1700V, while with the control DC voltage of 1750V the maximum absorption power is only 3.5MW. At the same braking distance of 4km, the maximum absorption power is 840kw with control DC voltage of 1700V, while with control DC voltage of 1750V the maximum absorption power is only 430kW.

3. Field operation performance

3.1. System configuration

A inverter-based utilization device of 2.5MW power rating is installed in a substation of Nanjing S8 line which is also equipped with the on-board resistor-based utilization device. As shown in Figure 6, the 2.5MW inverter-based device is installed in station B, and within the 5kM distance there are three other stations, namely StationA、StationC、StationD. As stated in the last section, the inverter-based device can recover the regenerative power from the braking train not only at the station B but also the adjacent stations.

Table 1 shows the recovered energy with different control DC voltages, and the lower control DC voltage yields more saved energy. Considering there should be enough margin to ensure the control DC voltage is higher than the floating voltage of diode rectifier device, so the control DC voltage is set to be 1730V for the field operation.

IOP Conf. Series: Earth and Environmental Science **133** (2018) 012008 doi:10.1088/1755-1315/133/1/012008

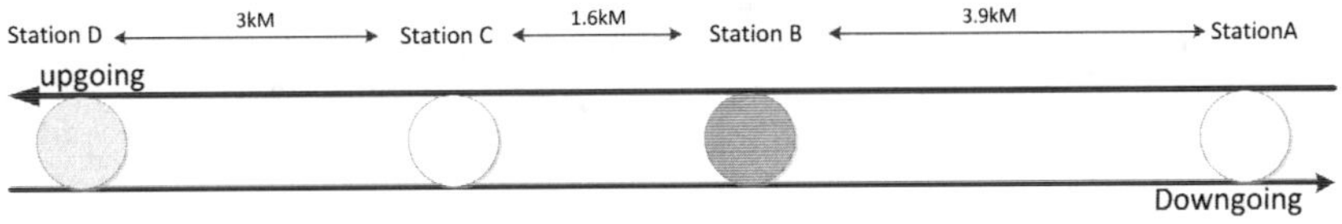

Figure 6. The adjacent stations condition of the inverter-based device

Table 1. Saved energy at different control DC voltage

Control DC voltage	Utilization energy per day(kWh)
1770V	2310
1750V	3115
1730V	4130
1710V	5355

3.2. The DC line voltage stabilization

The inverter-based utilization device can effectively reduce the fluctuation of the overhead DC line voltage. As shown in Figure 7(a), the overhead DC line voltage fluctuates greatly without the inverter-based device, when the train is braking the overhead DC line voltage rises to about 1800V and when the train is accelerating the overhead DC line voltage drops to about 1500V, while Figure 7(b) shows that the maximum DC line voltage decrease to about 1730V with the inverter-based utilization device, the inverter-based utilization device reduced DC voltage fluctuation.

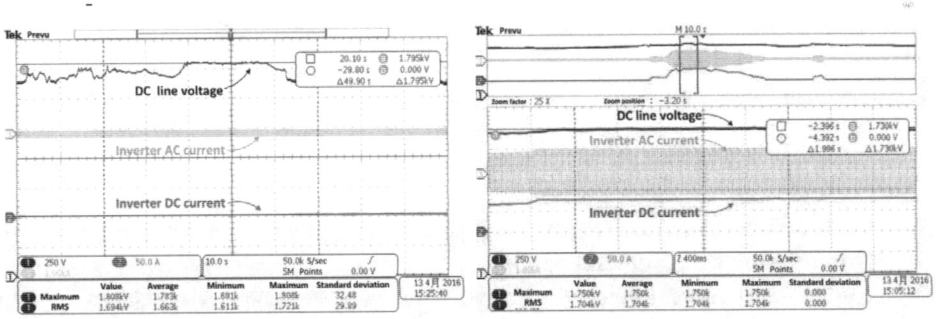

(a) voltage fluctuation without inverter (b) voltage fluctuation with inverter

Figure 7. The overhead DC line voltage stabilization

3.3. Energy saving performance

The energy recover performance is related with many operation factors like train model, braking speed, timetable, passenger onboard rate, and etc. In this case, a model B metro train with 2 motor cars and 2

IOP Conf. Series: Earth and Environmental Science **133** (2018) 012008 doi:10.1088/1755-1315/133/1/012008

trails is used in this line, the train braking speed is 80km/h, and on the work days there are 268 train departure per day while on the off days there are 228 train departure per day.

In one year operation the inverter-based utilization device with the control DC voltage of 1730V recovered more than 1.2 million kwh, and more than 100000 kwh per month averagely.The monthly and daily energy saving data are shown in Figure 8.

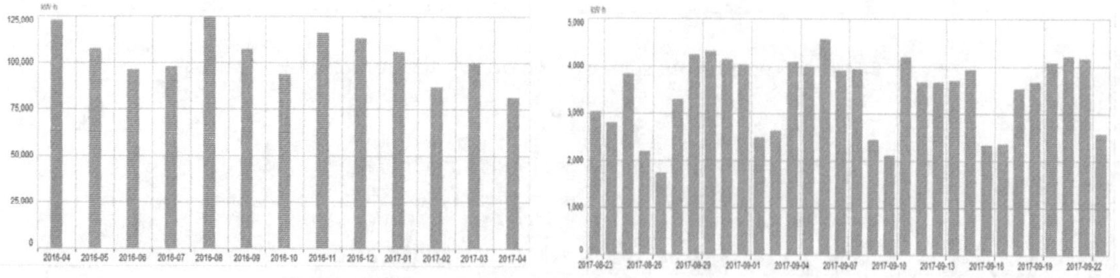

(a)Monthly data(kWh) (b) Daily data(kWh)

Figure 8. Energy saving data

4. Conclusion

The inverter-based regenerative power utilization device can effectively stabilize the overhead line DC voltage and recover the regenerative energy. Several key influencing factors like control DC voltage, braking distance are analyzed based on the absorption model, and the field data verified the lower control DC voltage can yield more saving energy. The field operation data of energy saving performance and the overheard DC voltage stabilization effect of an inverter-based power utilization device installed in NanJing S8 line are given, in the one year operation the device recovered more than 1.2 million kWh energy , at the same time the overhead DC voltage maximum value is constrained and reduced from 1800V to 1745V.

Reference

[1] Yang Jian, Huang Hou-ming, Fang yu. Energy Consumption of Trains on Shanghai Metro Line 2[J]. Diesel Locomotives, 2009,(4):23-25.

[2] LIU Baolin. Analysis on Energy Consumption of Metro Train[J]. Electric Locomotives and Mass Transit Vehicles. 2007,30(4):65-70.

[3] ZHOU Jian-bin, SU Jun, He Yong-bin. Utilization of Train's Regenerative Energy in Metro System [J]. Urban Metro Transit, 2004, 7(4): 33-35.

[4] ZHAO Li-feng, Zhang Fa-ming. Regenerative Energy Absorption Devices in Beijing Metro Line 5[J]. Modern Urban Transit, 2008, 2(1): 6-8.

[5] RUFER A, BARRADE P, HOTELLIER D. Power Electronics Interface for a Supercapacitor-based Energy Storage Substation in DC Transportation Networks [J]. European Power Electronics and Drives Journal, 2004, 14(4): 43-49.

[6] Yang G. Analysis of Urban Rail Transit Train Regenerative Braking Energy Feedback utilization Device[J]. Construction & Design for Engineering, 2016.

[7] Zeng Jianjun, Lin zhiming, Guo wanling. The regenerative braking energy absorption devices[J]. Electrified railway , 2008(6): 44-47.

IOP Conf. Series: Earth and Environmental Science **133** (2018) 012009 doi:10.1088/1755-1315/133/1/012009

Review on Water Distribution of Cooling Tower in Power Station

Zhang Huichao, Fang Lei, Guang Hao, and Niu Ying[1]

National Institute of Metrology, China. 18 Changchi Road, Changping, Beijing, China

Abstract. As the energy sources situation is becoming more and more severe, the importance of energy conservation and emissions reduction gets clearer. Since the optimization of water distribution system of cooling tower in power station can save a great amount of energy, the research of water distribution system gets more attention nowadays. This paper summarizes the development process of counter-flow type natural draft wet cooling tower and the water distribution system, and introduces the related domestic and international research situation. Combining the current situation, we come to the conclusion about the advantages and disadvantages of the several major water distribution modes, and analyze the problems of the existing water distribution ways in engineering application, furthermore, we put forward the direction of water distribution mode development on the basis knowledge of water distribution of cooling tower. Due to the water system can hardly be optimized again when it's built, choosing an appropriate water distribution mode according to actual condition seems to be more significant.

1. Introduction

Cooling tower is of great significant thermal equipment in power station which is helping releasing the heat that circulating cooling water absorbed in via interchanging between water and steam, so that the circumstance in condenser can maintain the standard of vacuum[1]. As a exothermic terminal equipment of low temperature, cooling tower is not only an important link to guarantee the stability of exhaust pressure, but affects vacuum degree in condenser by the temperature of the discharge water, therefore, cooling tower can release the heat of circulating water into air timely or not influences the economical efficiency of cold-end system directly. In practical, the improving vacuum of condenser can lead to the dropping of turbine exhaust temperature, which reduces the fuel consumption that generate equal quantity of load. The falling of cooling tower discharge water temperature can improve vacuum of condenser, which will influence cooling tower discharge water temperature and the coal consumption of power station.

At present, energy sources situation is becoming more and more severe, the state takes the issue of energy saving and emission reduction as the strategic key point of national economy development. The research on water distribution mode of natural draft cooling tower to optimize the water distribution system, on the one hand, can bring considerable economic benefit for the power station; on the other hand can save energy effectively.

2. Development of cooling tower

After entering electrified period, power plant produces great amount of waste heat in power generation process, the waste heat can hardly be further used according to the first law of thermodynamics and the second law of thermodynamics. Then the discussion about how to eliminate the waste heat pushes the cooling tower into rapid development. The principle of cooling water system is shown in figure 1 below:

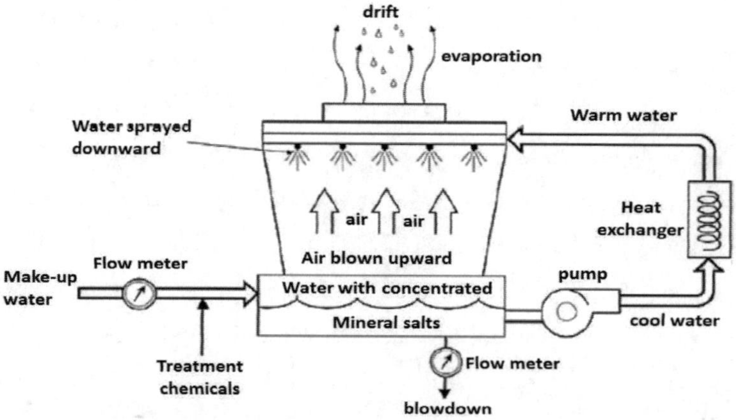

Figure 1. The principle of cooling water system

In 1904, the first industrial cooling tower was built in France, and since then, the research on the cooling tower mainly focus on the optimization of cooling tower performance. In 1920, the Netherlandish engineer F K Herson and G Kuypers completed the design of hyperbolic cooling tower. This kind of air cooling tower makes use of the temperature difference between the air around the tower and the air inside of the tower air to cooling, as the hot air inside the tower rises, the cold air is inhaled through air entrance in the bottom of the tower[2]. Consequently, such cooling tower don't need a fan, there circulates scarcely any hot air in the tower that will affect the performance, which forms very good thermal properties. In addition, the curvature exists in vertical and horizontal directions of tower wall, the stress that the lower part of tower wall subjected is lesser, which can reduce the wall thickness to save building materials; also the minimum vibration frequency of hyperbolic cooling tower wall is higher, which provides good dynamic characteristics. The earliest natural draft cooling tower (namely hyperbolic cooling tower) is built and put into use in British, and it's widely used since the 1930s; hyperbolic cooling tower group was built in northeast of China in 1940s. Natural draft counter-flow cooling tower is shown in figure 2.

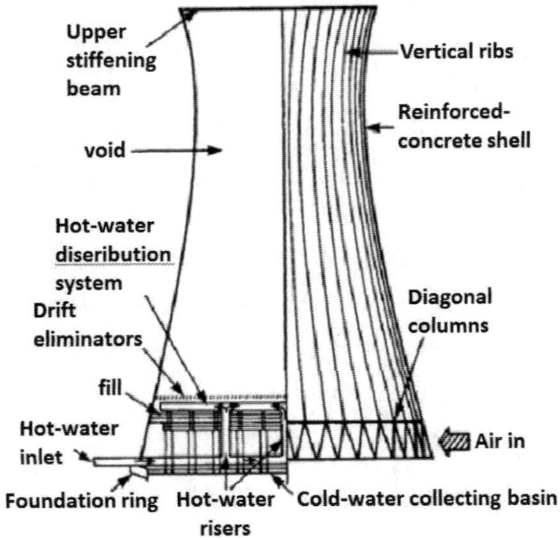

Figure 2. Natural draft counter-flow cooling tower

3. Analysis of water distribution mode

The role of the water distribution system is to ensure that hot water can be uniformly distributed on the packing surface within a certain limits, in order to make full use of the filler, which is an important aspect of improving the cooling effect of cooling tower. Besides, during the process of designing water distribution we should reduce Kinetic energy consumption as much as possible, pay attention to facilitating Maintenance Management and water flow adjustment, and minimize ventilation resistance. Water distribution mode of modern natural draft counter-flow wet cooling tower is classified main in two ways: piping distribution system and troughing distribution system. Piping water distribution system are divided into fixed and rotary two kinds.

3.1. Piping water distribution system

Piping water distribution system chooses steel pipes for water distribution, allocates the hot water to the various locations of the high elevation section of the cooling tower. The advantages of using this water distribution system is that flow velocity is high in the tube, and compared with the troughing water distribution system, this system takes up smaller area of the cooling tower ventilation, ventilating resistance is lower , and the water distribution is more uniform with the same water flow in the pipe. Due to high water pressure in the pipe, the flow of water through the splash device form small water droplets, which lead to good cooling effect, high quality of the installation, and difficult to grow algae. The disadvantages are that the splash devices require higher water supply pressure and higher water quality, poor quality of water will clog pipes, and the blockage is hard to clean.

3.1.1. Fixed water distribution. In fixed water distribution mode, pipe decorates mostly as a ring. The water pressure is relatively uniform and water distributes relatively uniformly with ringing arrangement. Especially when the pipe system is locally clogged, ring arrangement plays more superiority. Pressure losses in the piping water distribution system is small, and uniformity of the cooling tower water distribution will not be affected by small changes of each nozzle outlet pressure.

*3.1.2. Rotary water distribution.*In rotary water distribution, many channels perforate the water distribution pipes, the water rises from the tower center to the water distribution pipes forced by pressure, then outflows from the water distribution pipes, spays on the filler. Water distribution pipes are promoted to rotate in the opposite direction due to the reaction force When water outflow from the channels. Accompanied by the rotation of water distribution pipes, the water is periodically sprinkling on the filler. It's shown in figure 3：

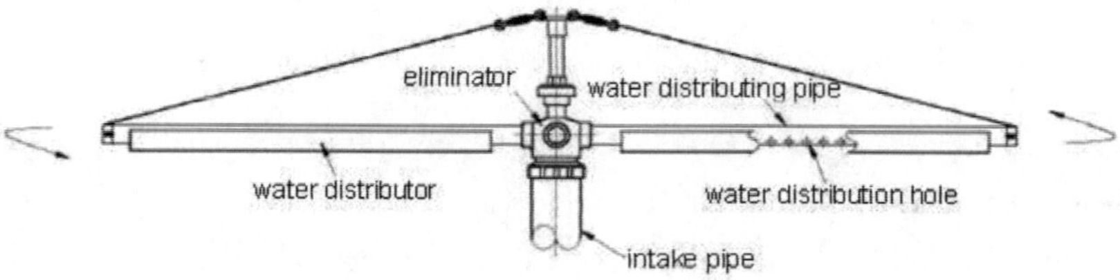

Figure 3. Rotary water distribution

3.2. Troughing Distribution System

Troughing distribution system can be divided into main channel, bypass channel and distributing channel three grades for water distribution, the channel section is a rectangle, and is made of reinforced concrete. The working status, hot water flows through the shaft with rectangular or circular cross-section into the main channel, then flows into bypass channel and distributing channel, finally spays on the packing through the shower installed on the distributing channel.

The advantage of troughing distribution system is that water pressure of supply system is low, which saves more water pump power consumption and can be easier cleaned compared with piping water distribution system. Defect is that algae and mud would be likely deposit on the water channel, and change the height of the flow, which aggravates corresponding areas packing load, and nonuniform packing load increased damage to packing; the mud can also lead to serious shower nozzle blocking, and reduce water distribution capability greatly; when the splash dish matched with nozzle plots a scale, water film becomes umbrellashaped and effect of nozzle splash becomes poor. What's more, the splashing water exists "hollow" phenomenon due to splash dish mud, namely water is splashing

IOP Conf. Series: Earth and Environmental Science **133** (2018) 012009 doi:10.1088/1755-1315/133/1/012009

around, but nozzle below forms an area without water. Meanwhile, structure of troughing distribution system is relative complex, the air resistance is high, and area without water often emerges in distributing channel under the situation of small flow of water, then the packing cannot be fully used.

3.3. Central shaft water distribution

In thermal power plant, in order to enable the cooling water temperature reaches the design value, and saving energy consumption (mainly the circulating water pump operation power consumption), and prevent cooling tower freezing in winter, natural ventilation cooling tower operation against type, use commonly the condition with inside and outside area water distributing, all summer with tower (inside and outside area are water distribution) operation, only the outside area with water distributing in winter operation. the central shaft water distribution combined with trough pipes mode appears in recent years, so that can realize water distribution in different partitions[3]. It's construction is shown in figure 4:

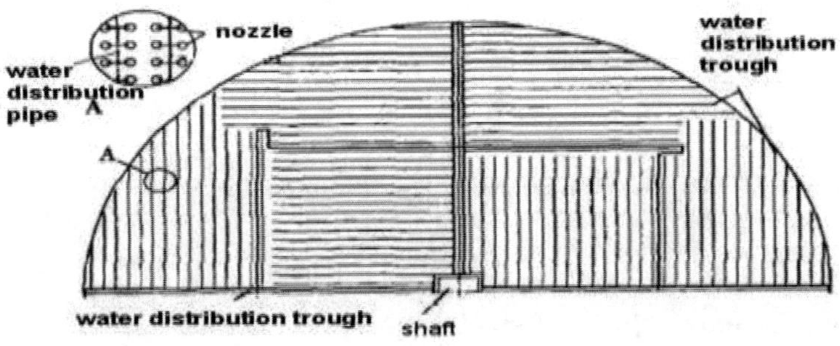

Figure 4. Central shaft water distribution combined with trough pipes

The advantages of this water distribution is energy saving, though flow velocity in pipes maybe so high that is far beyond the economic velocity range design code requires. The reason why flow velocity inside pipes is so high is that no matter how to decorate pipes the longest pipes increases and not less than $1/\sqrt{2}$ times of water tower shell radius with the increasing of cooling tower area. The longest water distributing pipe is so long that leads to high flow velocity.

3.4. Multiple shafts water distribution

Multiple shafts water distribution may have 3 or 4 or 5 shafts, each shaft lead branch from circulating water main pipe external the tower separately for water supply[4]. Because the layout, length and diameter of each branch pipe in the tower is different, the routing loss and local friction drag differs, can hardly allocated the total amount of water in different season of the year satisfyingly. Water distribution is not appropriate, the design of the cooling water distribution is difficult to achieve the expected requirements. The shaft with large water supply has high water level, and shaft with small water supply has low water level, the difference can reach about 0.5m according to the measure. The shaft with high water level pouring water density is large in water supply area, and the shaft with low water level pouring water density is small in water supply area, which leads to nonuniform of cooling tower water distribution, also affects cooling efficiency of cooling tower directly.

3.5. Suction Type Shaft Water Distribution

Suction type shaft water distribution is realized by siphon device in the shaft central, it's a kind of improvement of troughing water distribution system. The siphon device is composed of siphon mask and siphon weir. Suction type shaft water distribution is shown in figure 5:

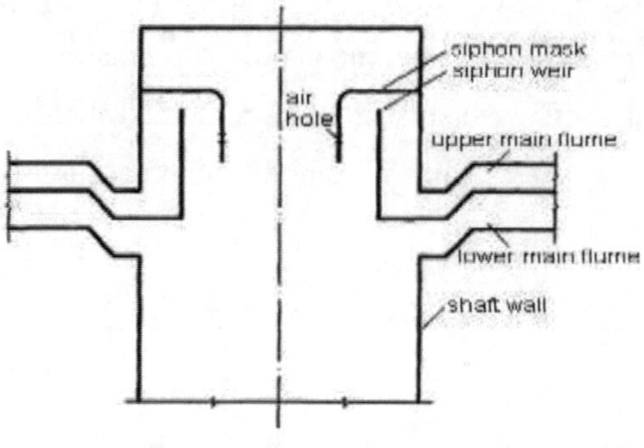

Figure 5. Suction type shaft water distribution

A suction type shaft is set in cooling tower central, a intake pipe without valve stepped into the tower feed water for shaft. The pressure main channels of inside and outside of cooling tower should be set respectively in order to realize the water distribution of tower area or only the outside area. Pressure main channel is divided into upper and lower layers, the upper for inside area, and the lower for outside area. The outside area is running water throughout the year, and the lower main channel connected with the shaft directly, the inside area can stop the water distribution in winter. As the upper main channel has water supply or not conditions, the siphon device is installed in the import. Since there is only one shaft water level, hydraulic head is the same for each main channel, and the hydraulic head for nozzle working is the same of the tower corresponding positions, so that water-spraying density is necessarily the same, which is helpful for improving the cooling efficiency of cooling tower. Suction type shaft water distribution has applied in cooling tower of many domestic projects[5]. The units capacity include unit level of 300 MW and 600 MW, such as two 8500m2 cooling tower in the Liaocheng power plant, two 9000m2 cooling tower in Dezhou power plant.

The advantages of suction type shaft water distribution are list as follows: it can automatic adjust the water distribution for all the area or only the outside area according to water quantity without valves in or out tower, and the nozzle working hydraulic heads of tower corresponding regions are the same lead to water distribution uniformity, and there do not exist water distribution uniformity in pipes, which can improve cooling tower efficiency. The disadvantage is that the siphon sometimes can't be created[6].

4. Conclusion

In the future, the optimization of water distribution system may carry out numerical simulations for cooling tower to calculate the temperature field, velocity field, stress field of the tower completely, at the same time, take the influence of the surrounding environment such as wind speed, the direction of the wind, temperature, humidity and so on into consideration, for the sake of getting closer to the actual engineering practice[7]. In addition, as the scope of optimization and adjustment is very small once the water distribution system is built, the power plant should pay more attention to choose the suitable water distribution system according to the cooling tower structure form. We need to pay attention to these main points: (1) Try as far as possible to uniformize the water distribution when decorating the shower nozzle. (2) The design should not only save pipe fitting but be beneficial to the pipeline hydraulic condition. (3) The cooling tower should be installed and maintained easily, and cleared conveniently.

The power plant operators should take the design documents as the basis, and combine with the operation regulations formulated with engineering practice of cooling tower, and be familiar with the operation maintenance requirements of cooling tower, grope operation rule and accumulate operation experience in practical operation.

References

[1] N. Williamson, S. Armfield, M. Behnia. Numerical simulation of flow in a natural draft wet cooling tower –The effect of radial thermofluid fields[J].Applied Thermal Engineering, 28(2008):178–189.

[2] R.V. Rao, V.K. Patel. Optimization of mechanical draft counter flow wet-cooling towerusing artificial bee colony algorithm[J].Energy Conversion and Management,52(2011):2611–2622

[3] PAN Chun, HAN Ling. Discussion on the Improvement of Water Distribution Uniformity of Cooling Towers[J]. INDUSTRIAL WATER & WASTEWATER, 31(2000):34-38.

[4] NIU Xiu-fu, SUN Feng-zhong, CHEN You-liang. Research and Development of Water Distribution of Natural Draft Counter-flow Wet Cooling Towers[J].Power System Engineering,24(2008):1-3.

[5] Zeng Jianzhu, Problems and Countermeasures in the Design, Construction and Operation of Suction Type Shaft Water distribution Cooling Tower[J]. Electric Power Construction, 28(2007):49-50

[6] Weng Xungan, Lu Zhenduo, Inquire into Suction Type Shaft Water Distribution for Natural Ventilation Reverse Flow Cooling Towers[J],Electric Power Construction,22(2001):9-12.

[7] JIN Tai, ZHANG Li, TANG Lei, YI Chao, GAO Xiang, LUO Kun, FAN Jianren. Three-dimensional Numerical Study on Water-Distribution Optimization in a Natural Draft Wet Cooling Tower[J].Proceedings of the CSEE,32(2012):9-15

IOP Conf. Series: Earth and Environmental Science **133** (2018) 012010 doi:10.1088/1755-1315/133/1/012010

Comprehensive evaluation of impacts of distributed generation integration in distribution network

Sujiang Peng[1], Erbiao Zhou[1], Fengkun Ji[1], Xinhui Cao[1], Lingshuang Liu[1], Zifa Liu[2], Xuyang Wang[2] and Xiaoyu Cai[2]

[1]State Power Economic Research Institute, 830000, Xinjiang, China
[2]School of Electrical and Electronic Engineering, North China Electric Power University, 102206,Beijing, China

Email: Xuyang Wang: 2567391970@qq.com

Abstract. All Distributed generation (DG) as the supplement to renewable energy centralized utilization, is becoming the focus of development direction of renewable energy utilization. With the increasing proportion of DG in distribution network, the network power structure, power flow distribution, operation plans and protection are affected to some extent. According to the main impacts of DG, a comprehensive evaluation model of distributed network with DG is proposed in this paper. A comprehensive evaluation index system including 7 aspects, along with their corresponding index calculation method is established for quantitative analysis. The indices under different access capacity of DG in distribution network are calculated based on the IEEE RBTS-Bus 6 system and the evaluation result is calculated by analytic hierarchy process (AHP). The proposed model and method are verified effective and validity through case study.

1. Introduction

On the situation that centralized development of renewable energy tends to saturate, distributed generation (DG) as an effective supplement is developed rapidly because of its low pollution, high energy efficiency and flexibility. By 2015, DG installed capacity in China has reached 45.33 GW. The DG capacity is expected to be 195 GW and the output account for 22% of the total production in 2020. Among varieties of DGs, there are several common use categories: wind turbine (WT), cell fuels, photovoltaic arrays (PV), micro gas turbine, small hydropower, etc. Because of DG connected to the network in a large scale, the economy, power quality, safety, and reliability of the distributed network are affected. In order to obtain more rational planning and better operation management that the distributed network needs under the effect of DG, it is essential to establish a scientific and systematic comprehensive evaluation Index system.

The previous evaluation method is mainly to evaluate single index. Considering the coordination coefficient among each index, a comprehensive evaluation index system is proposed in [1]. A reliability calculation method with consideration of the correlation between wind speed and load is proposed in [2]. In [3], the fuzzy AHP method, combined with the advantages of AHP and fuzzy evaluation method is applied to the distribution network comprehensive Assessment. The effect of DG on reliability, power quality, environment, economy of distributed network is taken into consideration in [4], and then an evaluation index system of DG affecting the network is established. The effects of DG different access modes on system reliability are studied in [5-6], In [5] DG's optimal access capacity and access point are studied from the aspects of inverse power constraint, voltage elevation and short-circuit current, and interval algorithm is used in [6] to compute the reliability Index of system before and after DG incorporation. Considering the output characteristics of wind power and PV system, the influence of DG on power flow distribution is studied in [7], and the variation performance of flow distribution is obtained by simulation. Compared with the previous work, the recent evaluation method has been improved greatly, and the evaluation index is more and more abundant, which increases the soft index of distribution network operation and management level, service, economy and so on.

In order to enrich the evaluation index system of distribution network, based on previous work, the effect of DG on power flow, power quality, line loss rate, reliability and operation management are synthetically considered in this paper and a comprehensive evaluation index system containing DGs is proposed, thus to provide scientific guidance to the planning and construction of the active distribution network and to strengthen the operation and management of distribution network.

2. Construction of comprehensive evaluation index system

2.1. Comprehensive evaluation index system

DG has attracted much importance due to their economy, flexibility, high efficiency and environmental friendliness. As more and more DGs are connected to the distribution network, the influences of the access of DG on distribution network are becoming more and more serious.

Significantly, the access of DG source changed the power supply structure of the distribution network, thus changing the power flow distribution, voltage distribution and so on. The small scale of distributed power supply and short construction period can reduce the construction cost of power grid. We can measure the economical efficiency of distributed power supply by the slow construction efficiency of the distribution network [8]. The single - power radiation network of distribution system is transformed into a variety of distributed power weak ring networks because of the access of distributed generation. The line loss rate is a comprehensive technical economic indicator that reflects the planning, design, technical equipment and economic operation level of power grid. The power loss of the grid depends on the power flow distribution of the distribution system. The access of distributed power supply makes the flow direction of each branch in the distribution network no longer one-way flow, so the access of distributed power supply will inevitably affect the power grid loss [9]. The line loss rate is related to the number of DGs, the capacity of the access point and the distribution of load. When the access capacity of the DG is greater than the load capacity of the system, the power grid loss will increase. Due to the randomness and instability of the output power of DG, the parallel operation of DG can cause voltage fluctuation, voltage flicker and harmonic pollution, etc., thus affecting the power quality of the power grid [10]. In order to ensure the good power supply quality, it is necessary

to integrate power quality into the comprehensive evaluation index system. When the system fails, the parallel operation of the DG will provide short-circuit current to the fault point. Different access modes and access capacity of DG have different effects on short-circuit current. The power supply reliability of the distribution network reflects the operating characteristics, the equipment structure, the running state and the power department's ability to manage the system and equipment of the distribution network [11-12]. As power supplies to power users in power outage, DGs can greatly reduce the user's annual power failure time. However, when analyzing the influence of DG on power supply reliability, the formation of isolated island and the randomness of DG output power should be taken into account. The radiant structure of the power supply of the distribution network has been fundamentally changed with the access of DG, reducing the sensitivity, speed, reliability and selectivity of relay protection, or causing the mal-operation and no-operation of protective devices [13-14]. The access of many kinds of distributed generation has a great influence on economic cost, voltage quality and safety of distribution network. Therefore, it is necessary to properly plan and improve the operation management level of the distribution network. In order to meet the above standards, a scientific and systematic comprehensive evaluation index system is urgently needed.
Therefore, it is necessary to quantify the several impacts of these DGs and make a comprehensive evaluation of the distribution network [15-16]. The economic and technical indexes of the influences of DG integration on the distribution networks are studied and analyzed in this paper. In this paper, the active power unit cost index is used to indicate the degree of delaying the investment of new equipment due to the access of DG; the grid loss index of distribution networks is used to show the economy of distribution network with DG; voltage pass rate index, total harmonic distortion rate index and power supply reliability index are used to represent the service quality of the distribution network with the integration of DG; the short-circuit capacity index and degree of influence of relay protection index are used to indicate the safety of the distribution network operation. A set of index system with seven indexes, which are active power unit cost, grid loss, voltage pass rate, total harmonic distortion rate, power supply reliability, short-circuit capacity and relay protection, is formed in the following figure.

2.2. Calculation methods of indexes

After introducing the integrated evaluation index system of distributed power access distribution network, next we will introduce the quantitative methods of each index in detail. Because the calculation methods of each index of the distribution networks with DG are different from that of the traditional distribution networks.

2.2.1. Active power unit cost

The cost of active power unit $C^{IRP}{}_{i,p}$ can be used to express the efficiency of power grid. If $C^{IRP}{}_{i,p}$ is negative, that indicates that the power of the transmission line of the node and the transformer is decreased, which reduces the use intensity of the existing equipment, thus delaying the investment of the power grid company in the new equipment. The calculation method of active power unit cost index ($C^{IRP}{}_{i,p}$) is as follows.

$$C_{i,p}^{IRP} = \frac{C_{i,p}}{\nabla p} \tag{1}$$

IOP Conf. Series: Earth and Environmental Science **133** (2018) 012010 doi:10.1088/1755-1315/133/1/012010

Where, $C_{i,p}$ is the cost of active power fluctuation of node i; Δp is the active power fluctuation value of node i.

2.2.2. Grid loss

The access of DGs can reduce the grid loss of distribution network. Because of the active power and reactive power provided by the distributed power source at the load point, the power of the flow on the line is reduced directly. The calculation method of grid loss (P_{loss}) s as follows [17-18].

$$P_{loss} = \sum R_i \frac{P_i^2 + Q_i^2}{U_i^2} \tag{2}$$

Where, P_{loss} is the total active loss of all branch roads; R_i is the resistance on the branch circuit i; P_i、Q_i and U_i respectively indicates active power, reactive power and the voltage amplitude of front-end node of branch ***i*.**

2.2.3. Voltage pass rate

Inappropriate access location and access capacity of DG may worse the voltage distribution of distribution networks. For instance, if DG is installed at the end node of the line, the voltage of the end node may exceed the normal range. According to distributed power supply different access mode, DGs can be divided into generator-based DG and converter-based DG. Different types of DG have different effects on power quality. The wind power generation is limited by the natural environment. The generation state of wind turbine is random, and the integration operating of wind generation can cause voltage instability and voltage fluctuation. In order to indicate the influence of the voltage distribution of the distribution networks with the access of DG, voltage pass rate (v_p) is calculated as follows.

$$v_{qr} = \frac{N_p}{N} \tag{3}$$

Where, N_p is the number of nodes with qualified voltage; N is the total number of nodes of distribution network.

2.2.4. Total harmonic distortion rate

Similarly, photovoltaic power generation, the micro gas turbine power generation is connected to the power grid through the inverter. The opening and closing of the converter switch circuit will result in the harmonic component of the switching frequency to be injected into the distribution network, and the harmonic distortion will occur in the voltage waveform. In this paper, the total harmonic distortion rate index (THD_u) is chosen to quantitative the impact degree of harmonic distortion, and the calculation method is as follows.

$$THDu = \frac{\sqrt{\sum_{h=2}^{H_{max}} U_h^2}}{U_1} \times 100\% \tag{4}$$

Where U_h is the RMS of the n_{th} harmonic voltage, U_1 is the RMS of fundamental voltage. H_{max} is the highest harmonics considered.

2.2.5. Power supply reliability

The power supply reliability calculation of distribution network with the access of DG is different from that of traditional distribution network. Due to the uncertainty and randomness of DG output power, the power supply reliability calculation method of distribution network with DG is complicated. In previous studies, the output power of DG is generally considered as constant value to calculate the power supply reliability. This method simplifies the computation process, but results in less accurate calculation of power supply reliability. In this paper, the output power of DG and load power adopt the time correlation model. The strategies used in the calculation are as follows. First, the output power of DG and load power at the moment of failure are calculated according to the time correlation model. Second, the power supply reliability calculation of distribution network with DG is carried out by considering the island operation mode of DG. The fault frequency and failure time of the load point are calculated by simulating the failure of components in the distribution network. The average service availability index (ASAI) is adopted in this paper. The calculation method of ASAI is as follows [19-20].

$$ASAI = 1 - \frac{\sum_{i \in R} U_i N_i}{\sum_{i \in R} 8760 N_i} \tag{5}$$

Where, U_i is the annual average duration of power failure of load point i; N_i is the number of users of the load point *i*.

2.2.6. Short-circuit capacity

The integration operating method of DG will provide short-circuit current to the fault point. When the access capacity of the DG reaches a certain degree, it may make the relay protection of the distribution network unable to make the correct action. The short-circuit capacity index S_c is selected to measure the security of distribution network with DG in this paper, and the calculation formula is as follows [21-22].

$$S_c = \sqrt{3} U I_f \tag{6}$$

Where, U is the bus voltage; I_f is the largest short circuit current.

2.2.7. Relay protection

The previous studies of the influence of DG access on relay protection of distribution network are studied and analyzed, and we find that the influences on the protection of fuse are mainly about two aspects. On the one hand, because of the three section current protection based on the circuit breaker of the traditional distribution network, the access of DG may destroy the coordination relationship of the setting value and setting time of protective elements. On the other hand, the reverse short-circuit current can cause the failure of the fuse, leading to the loss of reliability of relay protection [23-25].

In this paper, a three grade quantization model of relay protection is adopted [26]. When there is no DG access to distribution network, p_i=100. If the access of DG does no affect the normal operation of relay protection, p_i=90. If the access of DG affects the sensitivity of protection, but does not affect the correctness of the protection, p_i=75. Finally, if the access of DG leads to the mal-protection or refusal of protection, p_i=60.

IOP Conf. Series: Earth and Environmental Science **133** (2018) 012010 doi:10.1088/1755-1315/133/1/012010

3. Comprehensive evaluation method

3.1. Calculation method of index weight

Analytic hierarchy Process (AHP) is an effective method for quantitative analysis of non quantitative events and for objective analysis of subjective judgment. The basic idea of AHP is to decompose complex problems by establishing clear hierarchy structure. The calculations of indices which can describe the influence of DG on the distribution network are illustrated in section 2 and the weights of each index need to be determined in the comprehensive evaluation. The indices can not be compared directly because of the different factors such as dimension, quantity level and evaluation standard. The utility analysis method is adopted to obscure each index into a unified dimensionless index for comparison.

Based on the 1-9 reciprocal scaling theory [27], AHP compares each index and constructs the judgment matrix E. If the consistency test formula satisfies the formula (7), the consistency of matrix E can be accepted.

$$\mathrm{CR} = \frac{\lambda_{\max} - N}{(N-1)\mathrm{RI}} < 0.1 \tag{7}$$

Where: is the maximum eigenvalue of matrix E; N is the number of evaluation objectives, RI presents the average random consistency index. Then, the weight coefficient of evaluation objective x is computed by (8):

$$w_x = \frac{\prod_{y=1}^{N} e_{xy}}{\sum_{x=1}^{N}\prod_{y=1}^{N} e_{xy}^{\frac{1}{N}}}, 0 < w_x < 0.1 \tag{8}$$

Where e_{xy} represents the relative important value of index S_X to S_Y, w_x represents the weight coefficient of index.

3.2. Fuzzy Comprehensive evaluation method

Because of the difference of dimension, evaluation standard and nature of each index, it is unreasonable to compare each index directly. In this paper, the utility transformation method is used to get the index function to obscure the indexes, and to eliminate the factors such as the dimension of each index and the evaluation standard, then indices can be compared with each other under an unified dimension.

By using the weight coefficient and the utility value of each index, the comprehensive evaluation score of the scheme can be obtained, and the calculation formula is

$$S = \frac{100\sum_{x=1}^{N} w_x U_x}{M} \tag{9}$$

IOP Conf. Series: Earth and Environmental Science **133** (2018) 012010 doi:10.1088/1755-1315/133/1/012010

Where S represents the final score of a scheme; U_x represents the utility value of the index; set M=1; the minimum score of the scheme is 0, the maximum is 100.

4. Case Study

4.1. Case instruction

To verify the evaluation index and method proposed in this paper, the main feeder F4 of IEEE-RBTS BUS6 system is taken as an example to simulate different DGs integration. The system structure is shown in Figure1. The voltage base value of the test system is 10 KV, the convergence precision is 10-4, LGJ-150 conductor is used in the system, and there are 23 load points, 26 nodes, 30 lines, 14 breakers, 22 isolation switches and 13 fuses.

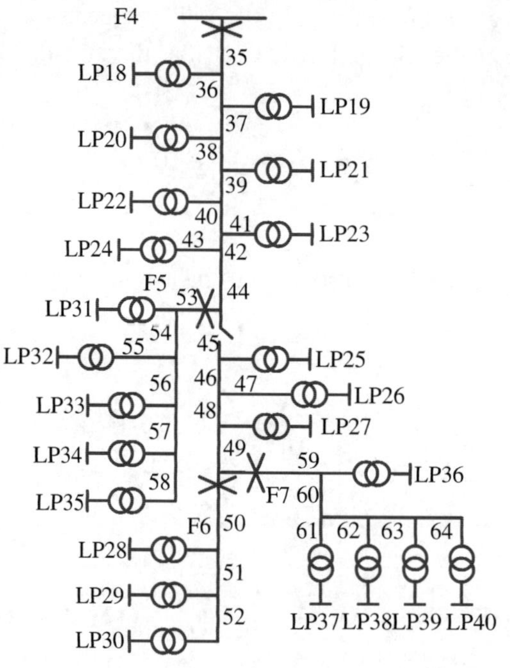

Figure 1. Test distribution network

Four scenarios of DG access plan are simulated in this paper. In Scenario 1, the access location and capacity of DGs are: LP30, 200kW; LP35, 200kW; LP40, 100kW. In Scenario 2, the access location and capacity of DGs are: LP22, 400kW; LP27, 400kW; LP30, 360kW; LP35, 360kW; LP40, 400kW. In Scenario 3, the access location and capacity of DGs are: LP22, 800kW; LP27, 900kW; LP30, 900kW; LP35, 900kW; LP40, 800kW. In Scenario 4, the access location and capacity of DGs are: LP22, 1000kW; LP25, 990kW; LP27, 1000kW; LP30, 990kW; LP35, 990kW; LP40, 1000kW.

Assuming that the system is completely reliable, the component failure rate data of the test system is shown in table.1 and the simulation results are shown in table.2.

Table 1. Equipment reliability data

Equipment	Fault rate	Fault time/h
line	0.050/km·a	4
transformer	0.015/a	200
breaker	0.002/a	4

Table 2. The Results of Different Indices

Index	Case 1	Case 2	Case 3	Case 4
Active power unit cost	4.48	4.14	3.98	3.91
Grid loss (MW)	0.79345	0.18695	0.16670	0.15880
Voltage pass rate	99.62%	99.98%	99.54%	99.24%
Total harmonic distortion rate	1.55%	1.62%	1.73%	1.82%
Power supply reliability	99.9465%	99.9787%	99.9811%	99.9816%
Short-circuit capacity	54.63	82.97	104.48	123.62
Relay protection	III	II	I	I

From the above table we can see that, the active power unit cost index and the grid loss index are getting lower as the access capacity of DGs increased. Thus, the integration of DGs improves the economic efficiency of distribution network. Because the randomness and instability of distributed power output power, with the increase of the access capacity of wind power and photovoltaic power generation, the influence of the output power of DGs on the voltage passing rate index of distribution network becomes bigger. So the voltage pass rate index presents a trend that increases first and decrease later. The opening and closing of the converter switch circuit of the converter-based DG will result in the switching frequency harmonic to be injected into the distribution network. Similarly, the total harmonic distortion rate index of distribution network with DG rises with the increase of the incorporation capacity of converter-based DGs.

Compared with scenario 1, the power supply reliability index of scenario 2 increases obviously. But the improvement degree of power supply reliability index of scenario 3 and scenario 4 is less obvious. When the power outrage occurs in the distribution system, DGs can operate in the isolated operation model, which ensures the reliability of the power supply for critical loads. So the power supply reliability of scenario 2, 3, 4 will increase. The reliability of distribution network is not only related to the access capacity of DG, but also to the components and network structure, so the increase of the index value of distribution network with higher access capacity is less obvious. The higher the access capacity of DG, the larger the short-circuit current and short-circuit capacity is provided. At the same time, the influence of relay protection on power distribution system is also increased. So measures to reduce the short-circuit capacity and the effect on relay protection should be adopted.

The score value of each index can be obtained by the calculation of nonlinear scoring function. The index weight calculated by AHP is as follows: 0.1160, 0.1465, 0.1081, 0.1081, 0.2193, 0.1563 and 0.1457. According to the index weights and scores, the comprehensive evaluation values of four scenarios are shown in table.3.

Table 3. Comprehensive evaluation values of different scenarios

Solution	Comprehensive Evaluation Value
Scenario 1	76.41
Scenario 2	87.23
Scenario 3	81.27
Scenario 4	79.38

From the comprehensive evaluation values of four scenarios, it can be seen that seven index values of distribution network with proper access capacity of DG are obviously improved. The comprehensive evaluation value of Scenario 2 is the highest, indicating that the comprehensive performance is the best. The access capacity of DGs of scenario 3 and scenario 4 is too large, which lead to serious impacts on the total harmonic distortion rate, short-circuit capacity and relay protection index of distribution network, so the improvement degree of comprehensive evaluation score is smaller than that of scenario 2.

5. Conclusion

(1) A comprehensive evaluation system fitting the practical demand of distribution network is proposed in this paper, which is based on the situation of various DGs integration, such as wind power and PV. The index system is composed of 7 indexes, active power unit cost, grid loss, voltage pass rate, total harmonic distortion rate, power supply reliability, short-circuit capacity and relay protection of distribution network. Through the calculation method of each index, a reasonable comprehensive score is obtained, and the result can be used as the basis of the network planning scheme.
(2) Different DGs integration to distribution network under proper access capacity can effectively improve the economic performance, voltage quality and safety index of distribution network, thus the comprehensive evaluation score of the distribution network can be improved. When the access capacity of DG is too high, some security and power quality problems will occur, such as ultra short circuit capacity, high harmonic distortion rate and large voltage fluctuation and so on.
(3) The comprehensive evaluation of future active power distribution network not only needs to consider the access of high penetration rate of DG, but also the new-type load, combined heat and power, and so on. It is necessary to reconsider the setting of the comprehensive evaluation index system and the calculation methods of the indexes, etc.

References

[1] Xue Zhenyu, Li Jingru, Chen Kai, et al A comprehensive evaluation method for distribution network considering the coordination coefficient *J. 2013 Power System Protection and Control* 41(**12**): 15-19.
[2] Sun Ruodi and Xie Kaigui, Reliability evaluation of distribution networks using Monte Carlo method considering correlations between wind speed and load *J. 2012 Power System Protection and Control* 40(**18**) : 12-18.
[3] Lin Jun, Wang Yi, Su Di. Application of an improved FAHP to the distribution planning *J.2008 High Voltage Engineering* 34 (**6**):1161-1167.

[4] Zhang Limei, Tang Wei, Zhao Yunjun, et al. The integrated evaluation of impact of distributed generation on distribution network *J.2010 Power System Protection and Control* 38(**21**): 132-135.

[5] Zhong Jiaqing, Ye Zhige, Lu Zhigang. Analysis of optimal allocation of penetration level and interconnected location of DG 2012 *J.Power System Protection and Control* 40(**7**): 50-55.

[6] Yu Kun, Chen Xingying, Chen Kai, et al. Study on the change rules of power flow in the urban power grid with distributed generation *J. 2013 Power System and Clean Energy* 29 (**2**) :5-10.

[7] Sui Xinxian, Wang Qian, Yang Yaqiang, et al. Study on impact of distributed generation on reliability of distribution network *J.2010 Sichuan Electric Power Technology* 33(**2**): 31-33, 52.

[8] Liang Huishi, CHENG Lin, SU Jian. Cost Benefit Analysis for Microgrid *J. 2011Proceedings of the CSEE*, 31(**Z1**).

[9] Wang J, Gao H, Zou G, Wu Z. Comprehensive evaluation of impacts of distributed generation on voltage and line loss in distribution network *C. 2016 International Conference on Electric Utility Deregulation and Restructuring and Power Technologies*. IEEE: 2063-67.

[10] Fan S, Pu T, Li L, Yu T, Yang Z, Gao B. Evaluation of impact of integrated distributed generation on distribution network based on time-series analysis *C. 2016 China International Conference on Electricity Distribution.* CICED.

[11] LU Jingjing, ZHAO Yuan, ZHAO Yongshuai, XIE Kaigui.A Point Estimation Method for Reliability Evaluation of Distribution Network With Distributed Generation *J. 2013 Power System Technology* (**8**):2250-57.

[12] GE Shaoyun, WANG Haoming. Reliability Evaluation of Distribution Networks Including Distributed Generations Based on System State Transition Sampling *J.2013 Automation of Electric Power Systems*, 37(**2**):28-35.

[13] ZHANG Yong. Review of Impact of Distributed Generation on Distribution System Protection *J.2010 Proceedings of the CSU-EPSA*, 22(**2**).

[14] Liu Lin Li Jianbing Liu Jingtao et al. Influence of Distributed Generation on Relay Protection and Reclose of Distribution Network *J. 2011 Sichuan Electric Power Technology*, 34(**1**):44-45, 79.

[15] Ochoa L F,Padilha-Feltrin A, Harrison G P. Evaluating distributed generation impacts with a multiobjective index *J. 2006 IEEE Transactions on Power Delivery*, 21(**3**):1452-1458.

[16] Chen C, WEN Y, LIU Z, Ren X, Zhang X. Comprehensive Evaluation Method of Distribution Network Including Various Types of Distributed Generation *J. 2015 Electric Power Construction*, 36(**1**):128-135.

[17] Cheng Zhanli. Research on the power flow calculation and optimized operation of distribution network with distributed generations *D. 2011 Southwest Jiaotong University.*

[18] Yan L, Xie Y, Xu J, Xue C, Zhao H, Bai L. Improved forward and backward substitution in calculation of power distribution network with distributed generation *J. 2013 Journal of Xi'an jiaotong university* 47(**06**):117-123.

[19] Qian K, Yuan Y, Zhou C. Study on impact of distributed generation on distribution system reliability *J. 2008 Power System Technology*, 32(**11**):74-78.

[20] Liu C, Zhang Y. Distribution network reliability considering distribution generation *J. 2007Automation of Electric Power Systems*, 31(**22**): 46-49.

[21] Yang S, Tong X, Liu J, et al. Short-Circuit Current Calculation of Distribution Network with Distributed Generation, *J. 2015 Power System Technology*, 39(**07**):1977-82.

[22] Wang C, Sun X. An improved method for short circuit calculation of distribution network with distributed generation *J. 2012 Automation of Electric Systems*, 36(**23**)**:**54-58.

[23] Wen Y, Wang X. Effect of distributed generation on protection in distribution system *J. 2008 Relay*, (**01**):12-14+26.

[24] Sun M, Yu J, Deng B. Analysis of impact of DGs on line protection of distribution networks *J. 2009 Power System Technology*, 33(**08**):104-107.

[25] Gao Y, Bi R, Yang W, Ding Y. Effect of distributed generation on relay protection in distribution system *J. 2009 Power System and Clean Energy*, 25(**04**):20-23.

[26] Xu B, Zhang H, Xian R. Evaluation method for impact of distributed electric resources on the relaying protection of distribution network *J. 2015 Electric Power Construction,* 36(**01**):142-147.

[27] Yan Ding，Long Yu，Cheng Haozhong et al. Power quality assessment for distribution network incorporation micro-grid in planning stage *J. 2013 Proceedings of the CSU-EPSA*, 25(**6**):9-15.

IOP Conf. Series: Earth and Environmental Science **133** (2018) 012011 doi:10.1088/1755-1315/133/1/012011

A Novel Degradation Identification Method for Wind Turbine Pitch System

GUO Hui-Dong

School of Electrical Engineering, Beijing Jiaotong University, Beijing 100044, China

Abstract. It's difficult for traditional threshold value method to identify degradation of operating equipment accurately. An novel degradation evaluation method suitable for wind turbine condition maintenance strategy implementation was proposed in this paper. Based on the analysis of typical variable-speed pitch-to-feather control principle and monitoring parameters for pitch system, a multi input multi output (MIMO) regression model was applied to pitch system, where wind speed, power generation regarding as input parameters, wheel rotation speed, pitch angle and motor driving currency for three blades as output parameters. Then, the difference between the on-line measurement and the calculated value from the MIMO regression model applying least square support vector machines (LSSVM) method was defined as the Observed Vector of the system. The Gaussian mixture model (GMM) was applied to fitting the distribution of the multi dimension Observed Vectors. Applying the model established, the Degradation Index was calculated using the SCADA data of a wind turbine damaged its pitch bearing retainer and rolling body, which illustrated the feasibility of the provided method.

1. Introduction

With the large scale of wind turbines and the construction of offshore wind farms, the equipment property loss, power loss and maintenance cost caused by wind turbine accidents will be greatly increased. The Condition Based Maintenance(CBM) of wind turbines can detect hidden troubles in advance, monitoring the development trend of faults, avoid early defects or hidden accidents resulting in more serious failure due to delayed maintenance actions, optimize unit maintenance plan, so as to save the cost of operation and maintenance of the turbines at last.

Condition based maintenance strategy based on Fault Diagnosis and Prognosis (PDF) technology has been extensively studied in the field of traditional power industry. Different with the traditional hydro and thermal power plant, small unit power rate and large number of wind turbines scattered installed in a large geographical area are the characteristics of wind farm. Huge number of sensors is needed to be installed in wind farms, large amount of data needed to be collected and processed, which lead to the cost of condition monitoring of wind turbines increase greatly. Data acquisition and monitoring system, which collects and records 10 min average values of meteorological, mechanical, electrical, temperature and other operating parameter is widely deployed in large wind turbines. The data acquired by the system contains rich unit status information, which makes it feasible to forecast the health status of unit and its main sub-equipment based on the Supervisory Control and Data Acquisition (SCADA) system. That means the condition based maintenance management of wind turbines can be implemented without adding sensors and wiring in turbines, which is an important significance to reduce the overall cost of operation and maintenance of wind farm [1].

Due to the intermittent wind resource, the output and the rotating speed of wind turbine change stochastically. The main shaft bearing, low speed shaft bearing of gear box, yaw and pitch bearing are low speed or incomplete rotation. These two factors limit the effect of vibration analysis method applied on the status identification of wind turbines. Because the main shaft bearings, yaw and pitch bearings are mostly lubricated by grease or grease and lubricating oil mixed lubrication, it is difficult to on-line monitoring the bearing status. When off-line analysis, it is difficult to ensure that the collected samples are involved in the lubrication work. Therefore, the effect of oil monitoring method for wind turbine status identification is also difficult to guarantee. In addition, threshold evaluation method using one parameter is difficult to adapt to the characteristics of wind turbine operation under variable conditions. Big data [2] and artificial intelligence algorithm has advantages in information correlation analysis, reasoning and optimization decision making, can be effectively applied in the faults detection of complex nonlinear systems [3]. Domestic and foreign scholars have carried out a lot of research on the application of artificial intelligence algorithm in FDP. As noted in [4], Artificial Neural Network (ANN) was applied to predict the wind turbine generation integrating wind speed and direction, air density, geographical location and other factors. Research in [5] indicated, the prediction method based on ANN can send warning information 2 days before the gearbox real fault occurs. Based on the principle of the first law of thermodynamics, a fault prediction model based on gear box temperature change data is established [6]. The application of rule-based expert system in fault diagnosis and prediction of wind turbines is studied in [7]. However, the fault identification sensitivity and accuracy are not enough, because most of the above studies directly use part of the operational data of wind turbines without considering the correlation and coupling within parameters.

An adaptive neuro-fuzzy inference system (ANIFS) was established in [8] to detect the early defect or potential fault for key wind turbine components, and the expert knowledge was used for Fault diagnosis and location. Time series model and Adaptive Neural Network are used to mine the features contained in historical data to achieve rapid detection of device abnormalities [9]. The effectiveness of SVM in fault diagnosis of wind turbine is verified by experiments in [10]. On the basis of principal component analysis (PCA) method to identify the operating condition subspace, health assessment model based on GMM multi state feature fusion was established in [11]. However, these methods can only predict failure a few minutes or one day before it occurs, which is helpless for conduct condition based maintenance strategy.

In this paper, an online identification model of wind turbine pitch system operating state based on regression model and GMM is established. On the basis of fusing multi-sensor data information of pitch system, the recognition accuracy of the deterioration state and the lead time of fault prediction are improved by the model established.

2. Introduction of wind turbine pitch system

2.1. Wind turbine pitch system structure and fault analysis

Pitch control system is an important part of wind turbine, because it not only ensures the effective utilization of wind energy by wind turbines, but also responsible for ensuring the wind turbine safety, avoid over speeding and falling tower accidents under the extreme bad situation. The pitch system of the wind turbine is installed at a height of tens of meters away from the ground, so it is difficult to maintain after the failure and the maintenance cost is high. Therefore, it is necessary and profitable to apply preventive maintenance on pitch system. The number of observation parameters of pitch system is large, which is convenient for data analysis. Considering the possibility of unit failure caused by failure of each subsystem of wind turbine and technical and economic feasibility of status detection of each subsystem, this paper choose pitch system as the key object of wind turbine condition monitoring. A single blade pitch device generally includes a controller, servo driver, servo motor, reducer, pitch bearing, sensor, angle limit switch, battery, transformer, power supply, etc. The common faults of pitch system can be divided into three types: executive mechanism failure, sensor fault and control system fault. Some of the faults such as leakage, which can be found by regular manual inspection, are not study in this paper.

2.2. Control strategy and operation characteristics of variable pitch system for wind turbines

The variable pitch variable-speed control strategy of wind turbine control wind energy conversion rate and power output by adjusting blade attack angle and rotating speed. In order to avoid the large power fluctuation and strong instantaneous stress load during the operation transition interval between the optimal blade tip speed ratio tracking control phase to the constant power control phase, Individual Pitch Control (IPC) technology and pitch stress reduction control strategy is widely adopted in wind turbines[12]. The advanced pitch control strategy, which can reduce vibration of blade and tower, improve the power generation efficiency and extend the turbine life simultaneously, require high reliability of pitch system.

The wind turbine actually operated in complex harsh weather environment, which lead to fatigue, overload, internal insulation aging and deterioration of the overall health of the equipment. As a result, the operation parameters of wind turbine have nonlinear relationship, the pattern of pitch system show diversity. Figure 1 shown the SCADA data 3D scattered point diagram for a typical operating wind turbine. Due to the dynamic changes of wind resources and the dynamic characteristics of wind turbines themselves, the distribution of the operating parameters of the pitch system in fig 1 is not completely consistent with the designed power curve.

2.3. SCADA operation data of variable pitch system

Variable pitch system is divided into two types according to different driving modes: hydraulic variable pitch and electric variable propeller. The monitoring parameters of typical electric pitch system include unit operation parameters, pitch motor parameters, pitch converter parameters, standby power parameters, etc. Totally, 9 feature vectors such as wind speed, active power, rotor speed, three blades pitch angle and pitch drive currencies were selected from 47 parameters recorded in SCADA to indicate the pitch system operation status. Despite the obviously inconsistent between actual power curve reflected by wind speed and active power and the designed power curve, which can be find by field operators easily, this paper try to excavate the deep deterioration symptom regarding the pitch system as a Multi Input Multi Output (MIMO) regression model, where wind speed, power generation regarding as input parameters, wheel rotation speed, pitch angle and motor driving currency for three blades as output parameters. Express characteristic vector of pitch system as formula below:

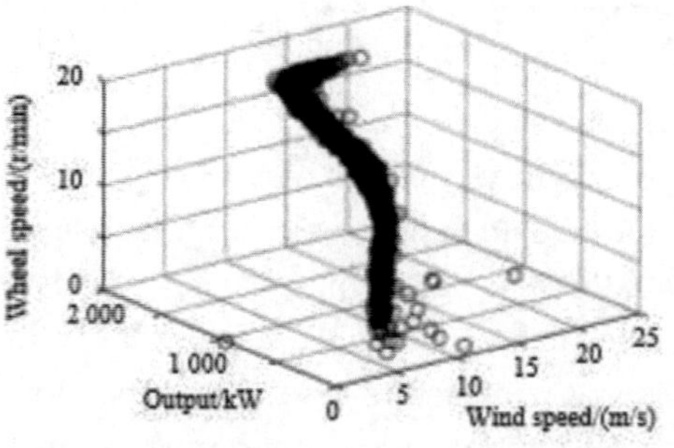

Figure 1. The SCADA data 3D scattered point diagram for a wind turbine.

$$X_N(M)=\begin{bmatrix} x_1(1) & x_1(2) & \cdots & x_1(M) \\ x_2(1) & x_2(2) & \cdots & x_2(M) \\ \vdots & \vdots & & \vdots \\ x_N(1) & x_N(2) & \cdots & x_N(M) \end{bmatrix} \tag{1}$$

Where, N is for sample size, M is for dimension of eigenvector.

3. LSSVM regression model of pitch system

Pitch system parameters regression has some difficulties. It is difficult to accurately describe the operation of pitch system through functions. A large amount of data from wind turbine SCADA system is transmitted and accumulated in the form of data stream, so the output of the regression model needs to be quickly obtained.

Support vector machine (SVM), for whom constructing training samples based on historical tag data is needed is a supervised learning method based on statistical learning theory developed in the middle of the 90s. SVM can improve the generalization ability of learning by seeking the minimum empirical risk and confidence risk, and obtain good statistical regularity under the condition of less training samples. If the sample data is linear non separable, SVM can use relaxation variables and kernel function to realize nonlinear classification. This paper applies the Least Squares Support Vector Machine, which transforms the optimization problem into a convex two programming problem with only equality constraints, training MIMO nonlinear model of pitch system.

Assume all training data $\{x_i, y_i\}$ can be map to high dimensional space using nonlinear mapping $\varphi(\cdot)$, regression of nonlinear functions in an input space can be converted as estimation of linear function in high dimensional space. The form of regression function is $y(x)=w^T\varphi(x)+b$,where w is weight vector, b is constant term.

$$\begin{cases} \min & \frac{1}{2}\|w\|^2 + \frac{C}{2}\sum_{i=1}^{n}\zeta_i^2 \\ \text{s.t.} & w^T\varphi_i(x_i) + b + \zeta_i = y_i \end{cases} \tag{2}$$

Where, relaxation variable $\zeta_i \geq 0$, $i = 1,\cdots,n$, C is penalty factor。 Lagrangian multiplier $\lambda \in R^{N\times 1}$,

By KKT rule, equations regarding $[b,\lambda]^T$ are obtained：

$$\begin{bmatrix} b \\ \lambda \end{bmatrix} = \begin{pmatrix} 0 & e1^T \\ e1 & \Omega + \frac{I}{C} \end{pmatrix}^{-1} \begin{bmatrix} 0 \\ y \end{bmatrix} \tag{3}$$

Where, $y = [y_1, y_2, \cdots, y_N]^T$, $e1 = [1,1,\cdots,1]^T$, $\lambda = [\lambda_1, \lambda_2, \cdots, \lambda_N]^T$, $\Omega_{ij} \in \varphi^T(x_i)\varphi(x_j) = K(x_i, x_j)$,

$i, j = 1,\cdots,N$。 $K(x_i, x_j)$ is the kernel function satisfying the Mercer condition. In this paper, we take radial basis function：

$$K(x_i, x_j) = \exp\left(-\frac{\|x_i - x_j\|^2}{\beta^2}\right) \tag{4}$$

solving the equations, achieve：

$$y(x) = \sum_{i=1}^{n} \lambda_i K(x_i, x) + b \tag{5}$$

The MIMO system is decomposed into several multiple input single output (MISO) subsystems Separately trained model by method in [13].

4. Condition identification of pitch system based on GMM

4.1. GMM application

Variable pitch system is divided into Define the deviation between the measured values of the system characteristic vectors and the regression results of the LSSVM as the "Observed Value", the Observed Vector of pitch system is:

$$\sigma_j = \left[\sqrt{(a_1 - a_1')^2}, \cdots, \sqrt{(a_m - a_m')^2}\right] \tag{6}$$

Where, a_i is the measured values of variable i of vector y, a_i' is the regression results of the LSSVM, m is the dimension of the vector。

In order to avoid the influence of environmental noise, sensor error containing the collected data, data pre-processing is needed by conventional time series correlation trend analysis. However, the data preprocessing process will cause loss of data information [14][15]. In this paper, the target detection method named "background subtraction" used in image processing was applied.

Due to its outstanding noise elimination and anti-interference ability, GMM algorithm regarding as an unsupervised learning data classification algorithm is widely used in image processing, pattern recognition. GMM uses Gauss probability density function describe data space distribution characteristics. Through increasing the number of Gauss probability density function, GMM can approach any continuous distribution by arbitrary precision. The method does not need to determine the prior probability distribution structure of the object, and can also avoid the difficulty of solving the nonlinear transcendental equations when estimating the parameters of the traditional probability density function. Hence, the Gaussian mixture model (GMM) was applied in this paper to fitting the distribution of the multi dimension Observed Vectors under normal conditions, and regard it as the background of the fault, then perceive the subtle changes in the distribution of observed values when deterioration of the system happened.

As equation below indicated, the probability of Observed Vector x_j at j time can be described as the weighted sums of probabilities that the value belongs to the K Gauss distributions:

$$p(x|\Theta)=\sum_{k=1}^{K} w_{k,j} N_k(x|\mu_k,C_k) \tag{7}$$

where, $N_k(x|\mu_k,C_k)$ is number k Gaussian distribution, μ_k is mean value, C_k is covariance; $w_{k,j}$ represent the prior probability of x_j belong to Gaussian distribution k, $\sum_{k=1}^{K} w_{k,j}=1$, $x=[x_1,x_2,\cdots x_m]^{\mathrm{T}}$; $\Theta=\{w_k,\mu_k,C_k\}$.

$$N_k(x|\mu_k,C_k)=\frac{1}{(2\pi)^{n/2}|C_k|^{1/2}} e^{-\frac{1}{2}(x_j-\mu_k)^T C_k^{-1}(x_j-\mu_k)} \tag{8}$$

4.2. Background model training

Likelihood function $L(\Theta|X)$ for sample $X=(x_1,\cdots,x_N\}$ is:

$$\log(L(\Theta|X))=\log[\prod_{j=1}^{N} p(x_j|\Theta)]=\sum_{j=1}^{N}\log p(x_j|\Theta)=\sum_{j=1}^{N}\log(\sum_{k=1}^{K} w_k N_k(x_j|\mu_k,C_k)) \tag{9}$$

Since the equation contain logarithm of sums, applying Expectation Maximization Algorithm (EM) for parameter estimation:

1) Initialization

Covariance matrix is unit matrix, $w_{k,j}=1/K$, mean value is the first Observed Vector, turn E-step.

IOP Conf. Series: Earth and Environmental Science **133** (2018) 012011 doi:10.1088/1755-1315/133/1/012011

2） E-step

The probability of new Observed Vector x_j belonging to the Gauss distribution k is,

$$p(k \mid x_j, \Phi^h) = \frac{w_k N_k(x_j, \Phi^h)}{\sum_{k=1}^{K} w_k N_k(x_j, \Phi^h)} \tag{10}$$

3） M-step

prior probability, matrixes updating：

$$w_{k,t-1} = \frac{\sum_{j=1}^{t-1} p(k \mid x_j, \Phi^h)}{t-1} \tag{11}$$

$$\mu_{k,t-1} = \frac{\sum_{j=1}^{t-1} x_j p(k \mid x_j, \Phi^h)}{\sum_{j=1}^{t-1} p(k \mid x_j, \Phi^h)} \tag{12}$$

$$C_{k,t-1} = \frac{\sum_{j=1}^{t-1} p(k \mid x_j, \Phi^h)(x_j - \mu_k)(x_j - \mu_k)^T}{\sum_{j=1}^{t-1} p(k \mid x_j, \Phi^h)} \tag{13}$$

4） Convergence criterion

Iteration step E and M, until

$$\left| \log(L(X \mid \Phi^h)) - \hat{\log}(L(X \mid \Phi)) \right| < \varepsilon \tag{14}$$

Calculate $\log(L(X \mid \Phi))$ as equation (10), $\hat{\log}(L(\Phi \mid X))$ is the result applying the updated Φ 。

5） Determine the background distribution

Sort the k distributions by value of $w_k / \|\sigma_k\|^2$, σ_k is standard deviation, the C distributions in front is background：

$$C = \arg\min_c \left(\sum_{k=1}^{c} w_k > T \right) \tag{15}$$

4.3. Discrimination of system health status

1） Determine whether the newly acquired observations x_t match with background Gauss distribution：

$$\begin{cases} d_{k,t}^T d_{k,t} < \lambda^2 \\ d_{k,t} = (\sigma_{k,t} I)^{-1}(x_t - \mu_{k,t}) \end{cases} \quad (16)$$

$\lambda \in [2,3]$.

2） If matching, updating the parameters by equation (12), (13), (14).
3） Calculate degradation index (DI)[16],

$$\text{DI}=\frac{1}{\alpha}\sum_{t=1}^{\alpha} NLLP_t \quad (17)$$

Where, $NLLP_t$=-ln$\left[\sum_{k=1}^{k} w_k N_k(x_t \,|\, k;\mu_k, C_k)\right]$, α is sliding window, which can suppress the influence of random factors, reduce misjudgement rate.

4） Turn to 1） for next observations x_t 。

5. example analysis

5.1. Parameter identification of regression model

In order to avoid the influence of power curtain due to transmission line limitation, the wind farm connected to East China Power Grid and close to the load centre was studied by this paper. There are totally 24 sets of 2MW variable speed pitch unit installed in the farm. The unit happened a 191# fault, which is that 2# blade pitch is not synchronized with the other two blades, at 19:25 November 20, 2014 was studied. Maintenance personnel found blade pitch bearing cage and rolling body damage, needed to replace the variable pitch bearing. Totally, 60 354 history data from SCADA was applied to train the model. Excluding data during period of maintenance, commissioning, manual start and stop, and all active power less than or equal to zero data, the rest data almost represented all operating conditions of the wind power generation. The fluctuation range of wind speed is 2.9-23.3m/s (rated wind speed is 12.5m/s), and the power fluctuation range is 0.1-1997kW.

Firstly, data normalization was conducted using a linear transformation function, and unified mapping to [0,1] interval：

$$\overline{x}_k = (x_k - x_{\min})/(x_{\max} - x_{\min}) \quad (18)$$

Where, x_k , $x_{\min}$, $x_{\max}$, $\overline{x}_k$ are measurement variable, the maximum input sample, the minimum input sample and the normalized value respectively.

Through parameter optimization, the parameters of the LSSVM model are $C = 35$, $\beta^2 = 10$. Observed Vector consisted of rotor speed, Pitch angle and current of pitch motor for 1#, 2# and 3# blade, consists of 7 dimensions, $m = 7$.

5.2. GMM background recognition

The prior probability T used to determine the background distribution in GMM model is difficult to be deduced by theory, the influence of different T values on the identification effect is evaluated by experimental method in this paper. Maintaining K constant as 16, changing T values as 0.5,0.6,0.7,0.8 for test. As the test results shown in table 1, when T is 0.5 or 0.6, misjudgement occurred due to the selected background distribution can't cover the whole normal operation conditions. Detection time with T value of 0.7 is earlier than the time with T value of 0.8, since sensitivity of recognition is reduced when T value is 0.8, and there are too many Gauss distributions in the background model.

Table 1. Degradation detection with different T value.

Number	T value	Date of detection	Time of detection	Judgment result
1	0.5	2014/8/21	10:30	Error detection
2	0.6	2014/8/30	23:10	Error detection
3	0.7	2014/11/11	8:20	Correct
4	0.8	2014/11/13	7:40	Correct

5.3. Evaluation of identification results of deterioration

Calculating the DI applying the method in 4.3 section. A large sliding window α would decrease sensitivity and the actual deterioration would not be confirmed in time. When α is 20, the time of Deterioration Index appeared obvious change is 8:20 November 11, 2014. When α is 10, the time of Deterioration Index appeared obvious change is 17:10 on November 9, 2014, which indicated the deterioration of the pitch system. Hence, α is 10 in this paper.

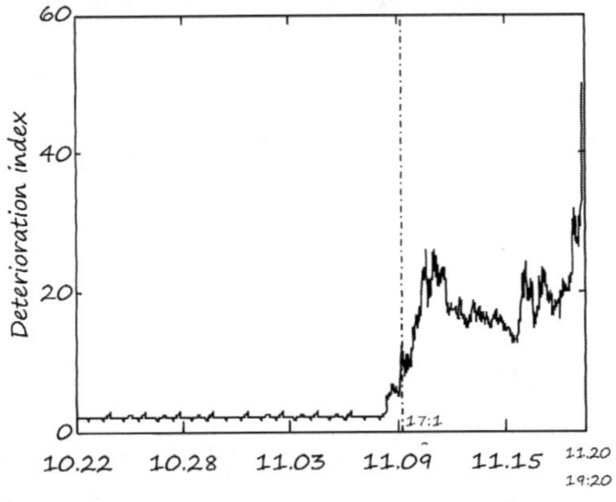

Figure 2. Degradation Index calculation result of wind turbine pitch control system.

As the DI curve in figure 2 indicated, Deterioration Index of the turbine began to increase on November 9, 2014, around 11 days ahead of the time when fault happened. Comparison between actual health state and model calculation results of wind turbines proved that this method has higher sensitivity to this kind of fault. The effectiveness of the method based on GMM proposed in this paper for the identification of the deterioration of the pitch control system is verified.

6. conclusion

Finding signs of potential failure in advance is the key to conduct CBM or kind of preventive maintenance for wind turbines, and improve economic benefits of the in-service wind farm. This paper proposed a novel method to identify the deterioration status of pitch system based on SCADA data.

1) The LSSVM regression model was established to solve the problem that the traditional pre-set threshold method could not adapt to the variable operating conditions of wind turbine。

2) The GMM model is used to fit the distribution law of the multi-dimensional observation values during normal operation of the unit, which avoids the shortcoming of information loss caused by the single parameter or the reduced dimension evaluation method.

3) The model application on a 2MW wind turbine with pitch bearing retainer and roller damage is presented in this paper. Compared with the traditional threshold method, the provided method can find the early characteristics of equipment degradation and significantly improve the lead time of fault prediction.

In order to facilitate the implementation of the preventive maintenance strategy for wind turbines, it is important to study the method of fault location and fault severity evaluation in the future works.

7. References

[1] Yang W. and Jiang J 2011Wind turbine condition monitoring and reliability analysis by SCADA information *Mechanic Automation and Control Engineering (MACE)* pp 1872-75

[2] YH Huang and XX Zhou 2015 Knowledge model for electric power big data based on ontology and semantic web *CSEE Journal of Power and Energy Systems* pp 19-27

[3] C Angeli and A Chatzinikolaou 2004 On-Line Fault Detection Techniques for Technical Systems: A Survey *International Journal of Computer Science & Applications* **1** pp 12-30.

[4] S Singh, TS Bhatti and DP Kothari 2007 Wind power estimation using artificial neural network. *Journal of Energy Engineering* **133** pp 46-52.

[5] MC Garcia and JD Pico 2006 SIMAP: Intelligent System for Predictive Maintenance: Application to the health condition monitoring of a windturbine gearbox *Computers in Industry* **57** pp 552-568.

[6] CS Gray and SJ Watson 2010 Physics of failure approach to wind turbine condition based maintenance *Wind Energy* **13** pp 395-405.

[7] Stuart R and Peter N 2009 *Artificial intelligence: a modern approach,* (New Jersey: Prentice Hall Upper Saddle River) p 898

[8] M Schlechtingen IF Santos and S Achiche 2014 Wind turbine condition monitoring based on SCADA data using normal behavior models: Part 1-system description *Applied Soft Computing* **14** pp 447-460

[9] Yan Yingjie, Sheng Gehao, Chen Yufeng, Jiang Xiuchen, Guo Zhihong and Du Xiuming 2015 An Method for Anomaly detection of state information of power equipment based on big data analysis *Proceedings of the CSEE* **35** pp 52-59

[10] An xueli, Zhao minghao, Jiang Dongxiang and Li shaohua 2011 Direct-drive wind turbine fault diagnosis based on support vector machine and multi-source information *Power System Technology* **35** pp 117-122

[11] Dong Yuliang, Li Yaqiong, Cao Haibin, He Chengbing and Gu Yujiong 2013 Real-time health condition evaluation on wind turbines based on operational condition recognition *Proceedings of the CSEE* **33** pp 88-95

[12] He Yulin, Liu Jun, Li Jun and Ren Haijun 2011 Variable-speed variable-pitch wind turbine control strategy optimization *Power System Protection and Control* **39** pp 55-60

[13] Zhou Xinran, Teng Zhaosheng and Zhao Xinwen 2009 Fast online system identification for MIMO using LSSVM *Journal of Computer Applications* **29** pp 2281-84

[14] Zhang Xiangrong, Jiao Licheng and Liu Fang 2008 Spectral clustering ensemble applied to SAR image segmentation *IEEE Transactions on Geoscience and Remote Sensing* **46** pp 2126-36.

[15] A lourenco and ALN FRED 2004 Comparison of combination methods using spectral clustering ensembles *Proceedings of the 4th International Workshop on Pattern Recognition in Information Systems* (Portugal：INSTICC Press) pp 222-233

[16] Yu Jianbo 2011 Bearing performance degradation assessment using locality preserving projections and Gaussian mixture models *Mechanical Systems and Signal Processing* **25** pp 2573-88

IOP Conf. Series: Earth and Environmental Science **133** (2018) 012012 doi:10.1088/1755-1315/133/1/012012

The wind power prediction research based on mind evolutionary algorithm

Ling Zhuang [1], Xinjian Zhao [1,2], Tianming Ji[3], Jingwen Miao[3] and Haina Cui[4*]

[1] State Grid Jiangsu Electric Power Company, Shanghai Road, Nanjing, 210000, China

[2] State Grid Nanjing Power Supply Company, Nanjing, 210000, China

[3] Information System Integration Company of Nari Group Cooperation, Nanjing, 211100, China

[4] Information System Integration Company of Nari Group Cooperation, Nanjing, 211100, China

cuihaina@sgepri.sgcc.com.cn

Abstract. When the wind power is connected to the power grid, its characteristics of fluctuation, intermittent and randomness will affect the stability of the power system. The wind power prediction can guarantee the power quality and reduce the operating cost of power system. There were some limitations in several traditional wind power prediction methods. On the basis, the wind power prediction method based on Mind Evolutionary Algorithm (MEA) is put forward and a prediction model is provided. The experimental results demonstrate that MEA performs efficiently in term of the wind power prediction. The MEA method has broad prospect of engineering application.

1. Introduction

The global energy crisis and global warming contributed to the rapid development of renewable energy. As a kind of renewable energy, wind energy has attracted the world's attention and affirmation[1-4]. Although wind power prediction can improve the safety and reliability of power system, the difficulty of power prediction always exists [5-6]. At present, the traditional methods of wind power prediction are as follows: time series analysis based on historical data, support vector machine (SVM)[7], neural networks, etc.

In paper[8], the auto-regressive moving average model was established by using the time series method to estimate the output of wind farm based on robust estimation. The prediction error is 5%. Qi Zhang et al. [9] introduced the present situation and related standard of wind power prediction, and optimized the wind power prediction model by neural networks. The performance of the proposed method was validated by simulation. In paper[10], a short-term wind power combination prediction method based on the overall mean empirical mode decomposition and improved Elman neural network was proposed. The simulation results indicate that this method can not only alleviate the influence of wind power's nonstationarity on prediction accuracy, but also avoid the modal aliasing problem caused by traditional methods. Mao et al. [11] proposed a wind speed and wind direction prediction model, and predicted the sub-power by ridgelet neural network.

In general, the time series method is used to predict the wind power by mining the data sequence, and it is suitable for short-term prediction with high prediction accuracy. Support vector machine method has changed the traditional risk minimization principle with strong generalization. Neural network method with strong robustness aims at the nonlinear characteristics of wind power. However, owing to the huge amount of training data, it is easy to fall into the local optimization.

On the basis, in this paper, the wind power prediction method based on Mind Evolutionary Algorithm (MEA) is put forward. This method optimizes the threshold value of the hidden layer weights by convergence and dissimilation, and greatly reduces the prediction error caused by the random generation of hidden layer weights' threshold. Based on the measured data, the wind power of a wind farm in China is predicted. The simulation results verify the validity of the model.

2. Wind power data preprocessing

In the process of wind power data acquisition, there may be a lack of wind power data. Therefore, we fill the loss data before carrying out the wind power prediction. In this paper, we choose a cubic spline interpolation function.

Suppose that (n+1) points were given:

$$a = x_0 < x_1 < x_2 \cdots < x_n = b \tag{1}$$

The cubic spline interpolation function *S*(*x*) is a cubic polynomial on each subinterval $[x_k, x_{k+1}]$, the m_k is calculated as:

$$m_k = S''(x_k) \tag{2}$$

set $x \in [x_k, x_{k+1}]$, the linear function pass (x_k, m_k) and (x_{k+1}, m_{k+1}) can be expressed as:

$$S''(x) = m_k \frac{x_{k+1} - x}{h_k} + m_{k+1} \frac{x - x_k}{h_k} \tag{3}$$

where $h_k = x_{k+1} - x_k$, According to the equation (3), the integral of function can be expressed as:

$$S(x) = m_k \frac{(x_{k+1} - x)^3}{6h_k} + m_{k+1} \frac{(x - x_k)^3}{6h_k} + [\frac{y_{k+1} - y_k}{h_k} - \frac{h_k}{6}(m_{k+1} - m_k)](x - x_k) + y_k - m_k \frac{h_k^2}{6} \tag{4}$$

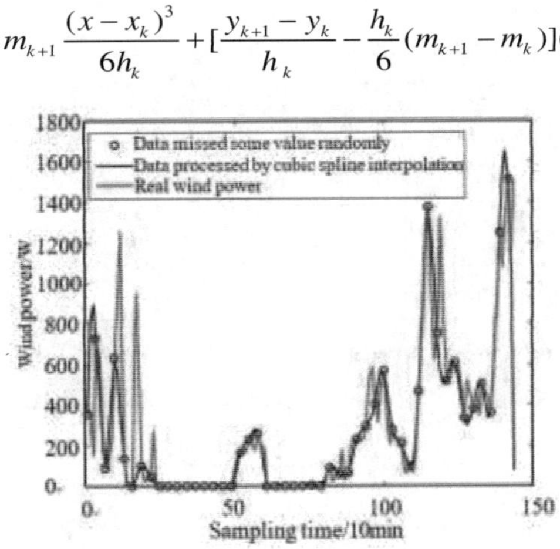

Figure 1. Results of the cubic spline interpolation fitting.

The main steps of the cubic spline interpolation are as follows:

(1) Determine the boundary conditions;

(2) Calculate the values according to the boundary conditions and form the equations;

(3) Solve the equations and obtain m_k ;

(4) Bring m_k into $S(x)$, and obtain an approximation of any point within [a, b].

Figure 1 displays the result of cubic spline interpolation fitting, and it's basically consistent with the real wind power data.

3. The wind power prediction model based on mind evolutionary algorithm

3.1. A brief introduction to mind evolutionary algorithm

The flow diagram of the MEA algorithm is given in Figure. 2.

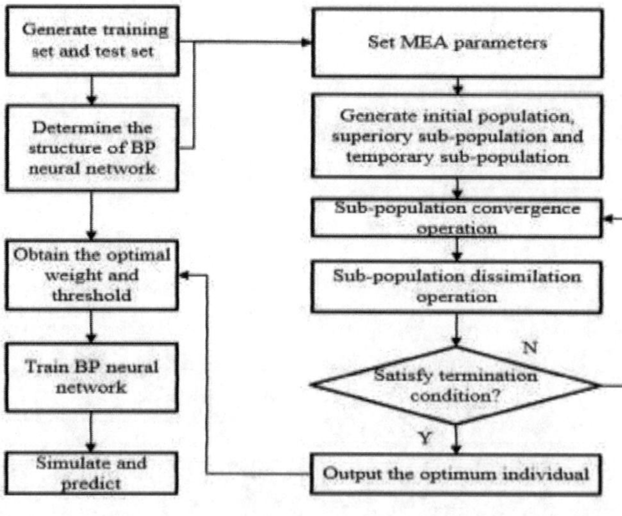

Figure 2. The flow diagram of the MEA.

Compared with the genetic algorithm, MEA has many characteristics:

(1) The group is divided into the superior sub-population and temporary sub-population, and carry on the convergence and dissimilation operation inside sub-population, convergence of semantic coordinate, but maintain their independence.

(2) Memorize more than one generation of evolutionary information that facilitates convergence and dissimilation.

(3) There are inherent parallelism in the structure of convergence and dissimilation operation

(4) Convergence and dissimilation operation can avoid the duality in the genetic algorithm, that is, it can produce good genes and also can destroy the original good genes.

3.2. Establish the wind power prediction model

Taking the measured wind speed , ambient temperature and wind power of a fan as an example, after data preprocessing, wind speed and ambient temperature are used as input to the wind power prediction model, then construct the eigenvector , the normalized eigenvector is calculated as:

$$E^{'} = [\frac{V_i}{\sqrt{V_i^2 + T_i^2}}, \frac{T_i}{\sqrt{V_i^2 + T_i^2}}] \tag{5}$$

Figure 3. The wind power prediction model based on MEA algorithm.

The wind power prediction model based on mind evolutionary algorithm is illustrated in Figure. 3.

4. Simulation and analysis

The simulation selects wind power measured data of a wind farm, and the data acquisition period of the wind farm is 10 min, the single fan capacity is 1500KW. In order to ensure the diversity of training samples, 1900 sets of experimental data as training samples from the collected wind farm data is selected randomly to train the MEA wind power prediction model. Then 100 sets of data as test samples in the remaining data for model validation is selected randomly. The other parameters of the MEA are shown in Table 1.

Table 1. Parameters of MEA.

population /size	superior sub-population /size	temporary sub-population s/size	hidden layer / size	iteration /times
200	5	5	5	10

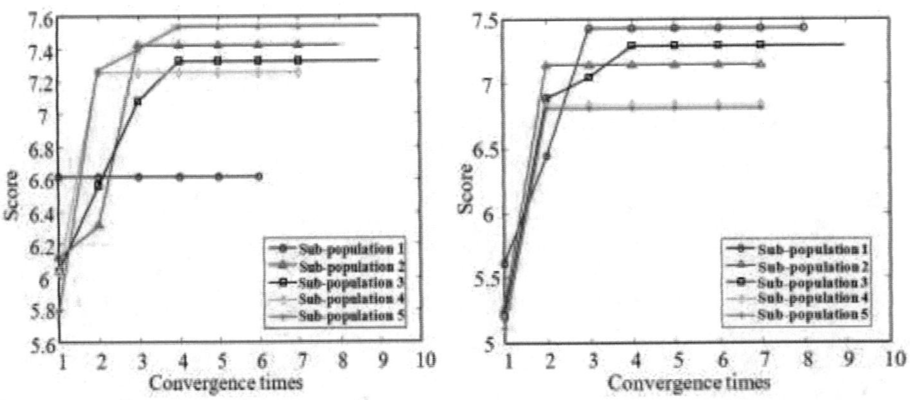

(a) The results of superior sub-population Convergence operation

(b) The results of temporary sub-population Convergence operation

Figure 4. The first convergence operation results of wind power prediction.

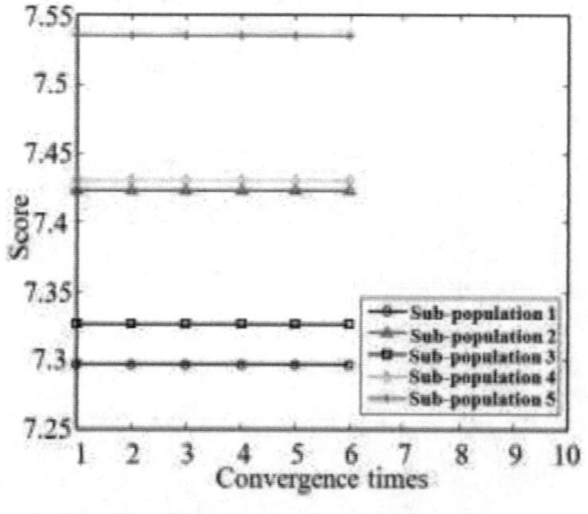

(a) The results of superior sub-population convergence operation

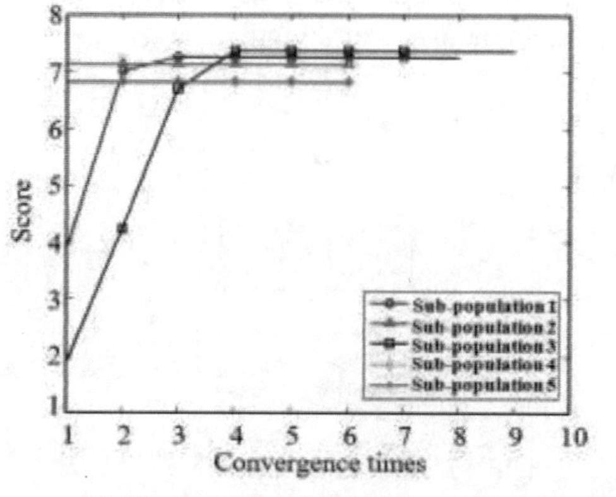

(b) The results of temporary sub-population convergence operation

Figure 5. The second convergence operation results of wind power prediction.

Figure 4-6 show the first convergence, the second convergence, and the third convergence operation results of the prediction model obtained by MEA. When the score of the temporary sub-population is lower than the score of the superior sub-population, the convergence operation is stopped and vice versa. As shown in Figure 4, the score of the temporary sub-population reach up to 7.4, which was much higher than that of the superior sub-population. Therefore, the MEA performs a dissimilation operation and then proceeds to the next convergence operation.

Based on the results given in Figure 6, all the scores of temporary sub-population are below that of the superior sub-population, therefore, the convergence operation stopped, and the wind power prediction model obtained by MEA achieves the optimal structure.

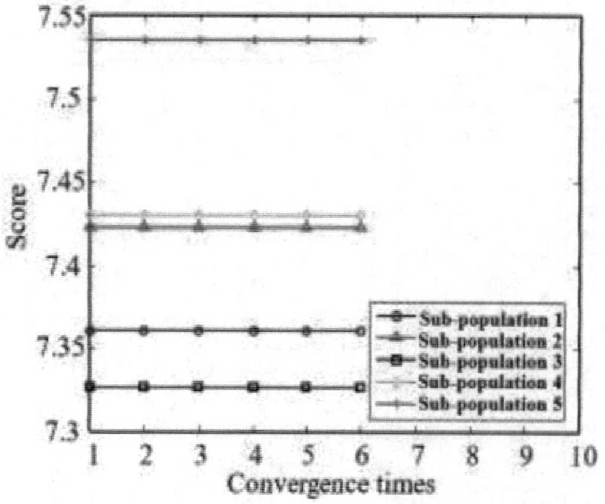

(a) The results of superior sub-population convergence operation

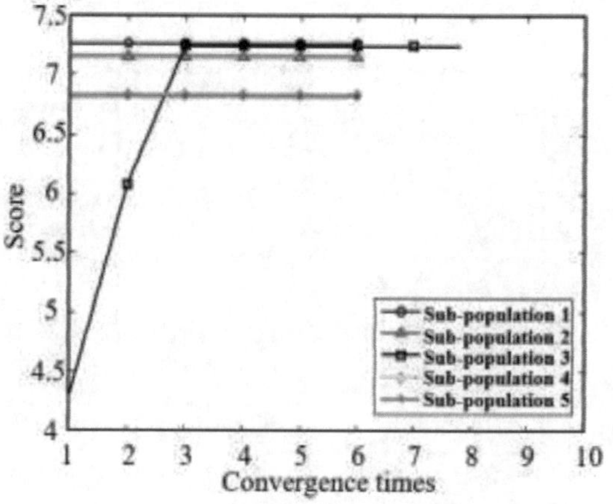

(b) The results of temporary sub-population convergence operation

Figure 6. The third convergence operation results of wind power prediction.

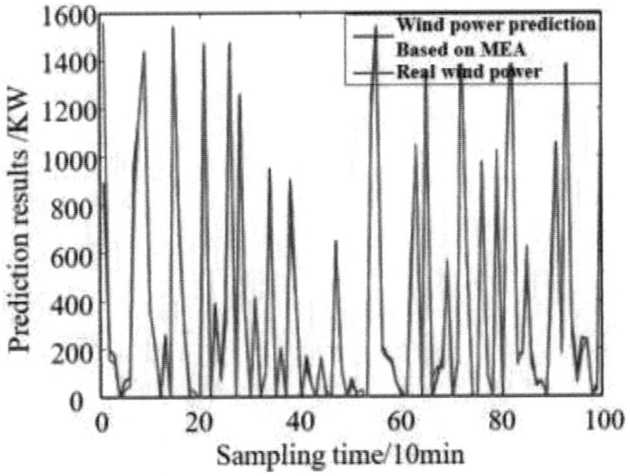

Figure 7. The results of wind power prediction Based on MEA.

With the model trained before, the wind power prediction simulation is carried out by using Matlab, the prediction results are given in Figure 7. The predicted result of MEA is basically consistent with the actual wind power.

Figure.8 displays the prediction errors based on MEA with sampling time varied from 0 to 100min. the maximum error is 16%, which satisfy the requirement of the maximum error can't exceed 25%. It shows that the mind evolutionary algorithm has high prediction accuracy and satisfy the engineering demand.

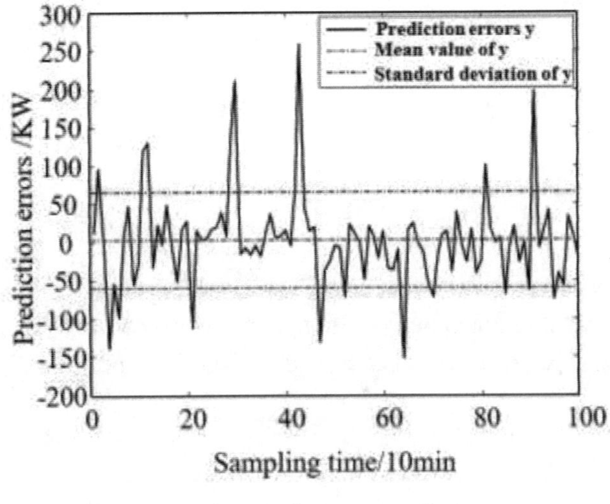

Figure 8. The wind power prediction errors.

5. Conclusion and Prospect

In this paper, the wind power prediction method based on Mind Evolutionary Algorithm (MEA) is put forward and the prediction model is provided. The eigenvector consists of ambient temperature and wind speed. The simulation results indicate that the proposed model has high prediction accuracy and laid the foundation of wind power prediction for the whole wind farm. In the future work, we can study the influence of the eigenvector and training sample on the prediction accuracy.

Acknowledgments
This work is supported by Science & Technology Program of State Grid Jiangsu Electric Power Company (Research on Key Technology of Substation Application Based on WIFI Cloud Platform).

6. References

[1] Landberg L, Myllerup L, Rathmann O, Petersen EL, Jorgensen BH, Badger J, et al. Wind resource estimation - An overview. Wind Energy. 2003;6(3):261-71.
[2] Yang M, Xiong H, Yan G, Mu G. Real-time prediction of wind power based on data mining and fuzzy clustering. Power System Protection and Control. 2013;41(1):1-6.
[3] Songlin Z, Meiqin MAO, Jianhui SU. Prediction of Wind Power Based on Principal Component Analysis and Artificial Neural Network. Power System Technology. 2011;35(9):128-32.
[4] Hongtao SHI, Jingling Y, Maosheng D, Jinmei W. A Short-term Wind Power Prediction Method Based on Wavelet Decomposition and BP Neural Network. Automation of Electric Power Systems. 2011;35(16):44-8.
[5] Ramirez-Rosado IJ, Fernandez-Jimenez LA. A new electric power forecasting model for wind farms. Proceedings of the 23rd IASTED International Conference on Modelling, Identification, and Control. 2004:398-401.
[6] Foley AM, Leahy PG, Marvuglia A, McKeogh EJ. Current methods and advances in forecasting of wind power generation. Renewable Energy. 2012;37(1):1-8.
[7] Mohandes MA, Halawani TO, Rehman S, Hussain AA. Support vector machines for wind speed prediction. Renewable Energy. 2004;29(6):939-47.
[8] Yang X, Sun H. Wind Speed Prediction in Wind Farms Based on Time Series Model. Journal of Power Engineering. 2011;31(3):203-8.
[9] Filik UB, Filik T. Wind Speed Prediction Using Artificial Neural Networks Based on Multiple Local Measurements in Eskisehir. 3rd International Conference on Energy and Environment Research, Iceer 2016. 2017;107:264-9.
[10] Wang H, Hu Z, Chen Z, Zhang M, He J, Li C. A Hybrid Model for Wind Power Forecasting Based on Ensemble Empirical Mode Decomposition and Wavelet Neural Networks. Transactions of China Electrotechnical Society. 2013;28(9):137-44.
[11] Amjady N, Keynia F, Zareipour H. Short-term wind power forecasting using ridgelet neural network. Electric Power Systems Research. 2011;81(12):2099-107.

IOP Conf. Series: Earth and Environmental Science **133** (2018) 012013 doi:10.1088/1755-1315/133/1/012013

Small Hydropower Development in Rwanda: Trends, Opportunities and Challenges

Gasore Geoffrey[1], Daniel Zimmerle[2] and Etienne Ntagwirumugara[1]

[1]Africa center of Excellency in Energy for Sustainable Development, University of Rwanda, Kigali-Rwanda,gasoreg60@yahoo.com, etienne.ntagwirumugara@gmail.com
[2]Colorado State University, Colorado, USA, Dan.Zimmerle@colostate.edu

Abstract. The Rift Valley region of Sub-Saharan Africa represents a promising area for the development of small (<5MW) hydropower resources. This study compiles data from government and UN agency reports to analyze different development outlooks. The study found that there has been a rapid deployment of small hydropower in the last 10 years. From the current total deployed small hydro of 47.5 MW, 16.5MW (35%) were deployed from 1957 to 1984 while the remaining 31 MW (65%) were deployed from 2007 to 2017. While all systems constructed prior to 1985 are grid-connected, one third of the 24 facilities constructed after 2007 are connected to off-grid systems. The study provides an overview of the economic incentives for developing small hydropower systems in Rwanda and the potential contribution of that development to Rwanda's electrification goals.

1. Introduction

Globally, the most-used definition of small hydropower is hydropower units with a rated capacity of 10 MW or less, but many countries define their own classification to meet local needs [1]. The Sector guidelines for environmental impact assessment (EIA) for Hydroelectricity Development in Rwanda by Rwanda Environmental Management Authority, classified Small hydropower plants by their capacity as: Mini hydro ranging from 500kW-10MW, Micro hydro 5-500kW and Pico Hydro which is less than 5 kW [2].

Small hydro power plants can also be classified according to head where ultra-low head of less than 3m;low-head of 3-40 m head; and medium-to-high head with more than 40 m head [3]. A recent government study estimates that Rwanda has 333 potential sites for small hydropower development and recommends medium- to high-head pico-and micro-hydro, using run-of-river plant designs [4].

According to hydropower atlas of 2007 [5], the majority of Rwanda potential hydropower sites have a capacity of 50-1000 kW of electrical production. Figure1 illustrates a small run-of-river hydropower facility designed for 300 kW. This type of facility is typical of both the location and architecture expected for many hydropower facilities in Rwanda. An upper catchment (left panel) slows the stream flow and allows some portion of the entrained silt to settle out. Design and maintenance of the silting basins are of paramount importance in Rwanda, new to the high silt load in many streams. Water is conveyed through an open channel (upper left in left photo) for some distance (approximately 2 km in this example) along the valley wall before emptying into 0.5 m diameter penstock that drops 30 m to the turbine house.

Impulse or cross-flow turbine types are frequently utilized for the heads seen in Rwanda. The example turbine is a gated cross-flow type (blue, right photo) driving a constant-speed synchronous generator (orange). Generation is governed by gating water flow (gray valve arms). Ingestion of silt into the turbine equipment presents significant issues for small Rwandan hydropower facilities, and may cause significant wear on both turbine blades and bearing seals.

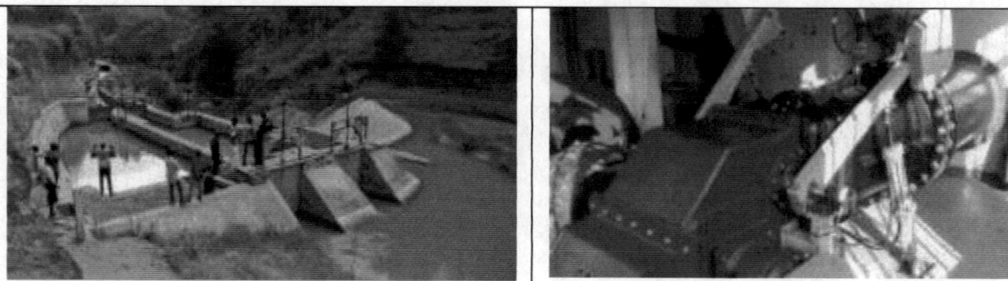

(a) Upper holding reservoir and silt removal basin.

(b) Penstock (green), control gate (black), cross flow turbine (blue), and generator (orange).

Figure 1. Example of small run-of-river hydropower facility in Rwanda.

Hydropower is the main source of electricity generation in Rwanda. Out of a current energy generation capacity of 210 MW, renewable resources represent 53.7% of the total energy generation and hydro power contributes 48% [6]. Fig.2 illustrates the domestic electricity generation in Rwanda.

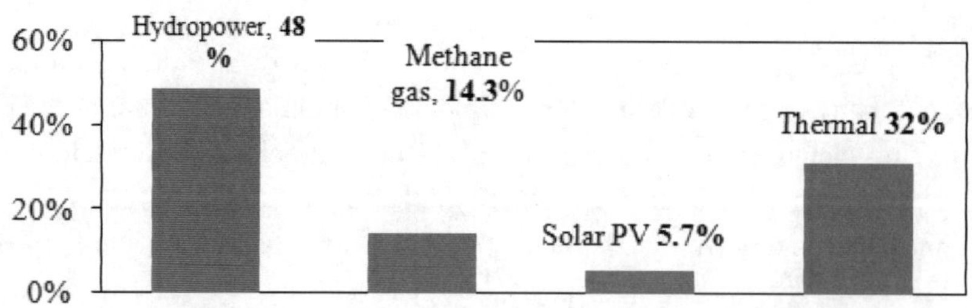

Figure 2. Resource mix for Rwandan Current energy generation.

In Rwanda's electrification strategy targets 512MW installed power generation capacity by 2023/24 and universal access to electricity for all households by 2023/24[6, 7]. Current electrification data indicate that electrification has reached 40.5% of the population, of which 29.5 % are grid connected and 11% are serviced by off-grid systems [6]. Under Rwanda rural electrification strategy, mini-grids will be developed by the private sector with government assistance in identifying potential sites and establishing framework through which they can become financially viable investments. Given the distributed nature of hydropower locations, the high targets for off-grid development will likely encourage small hydropower development.

In Rwanda rural electrification strategy, the levels of access are defined in five multi-tiers as illustrated in Table 1. The government plans to finance rural electrification in partnership with private sector.

IOP Conf. Series: Earth and Environmental Science **133** (2018) 012013 doi:10.1088/1755-1315/133/1/012013

The government will provide a basic level of electricity access to those with the lowest income by providing Tier 1(see Table 1) access to households with the lowest incomes and will help private sector reduce risks by providing of a risk-mitigation facility for future projects. The government has also established a policy and regulatory framework to incentivize the private sector to recycle redundant items such as kerosene lamps and used batteries [7].

Rwanda has electrified faster than neighbouring east African member state as illustrated Table 2, catching up to neighbouring Tanzania and Kenya in 2013.

The remainder of this paper will discuss the development environment for small hydropower in Rwanda. Overall development trends are discussed in Section 2, followed by a discussion of factors that have encouraged development, current challenges and opportunities. Finally, Section 6 provides conclusions and policy recommendations.

Table 1. Energy usage levels in Rwanda grouped in tiers.

Level	Tier 1	Tier 2	Tier 3	Tier 4	Tier 5
Energy usage	Household lighting, Radio and phone charging	Household lighting, Radio, phone charging and basic appliances (TV or fan)	Household lighting, Radio, phone charging, basic appliances (TV or fan), medium appliance such as low power refrigeration	Household lighting, Radio, phone charging, basic appliances (TV or fan), medium appliance such as low power refrigeration, high power appliances such as pumping.	High power suited to commercial and industrial uses.

Source: [7]

Table 2. Comparison of Electrification rate for the five East-African Countries.

Country	1990	2000	2010	2013
Burundi	0%	4%	5%	5%
Kenya	11%	15%	18%	20%
Rwanda	2%	6%	11%	21%
Tanzania	5%	9%	15%	24%
Uganda	7%	9%	9%	15

Source [8]

2. Development Trends

2.1. Historical Development

There has been rapid development in small hydropower in Rwanda since 2007. A full list of operational small hydropower facilities, at least as known to the government, is provided in Appendix1.These facilities account for an installed capacity of 47.5 MW. No small hydropower facilities were built between 1985 and 2006. Facilities built prior to 1985 were more traditional developments, larger – averaging 16.5 MW/plant – and fewer – 4 facilities – than recent development (Fig 3).

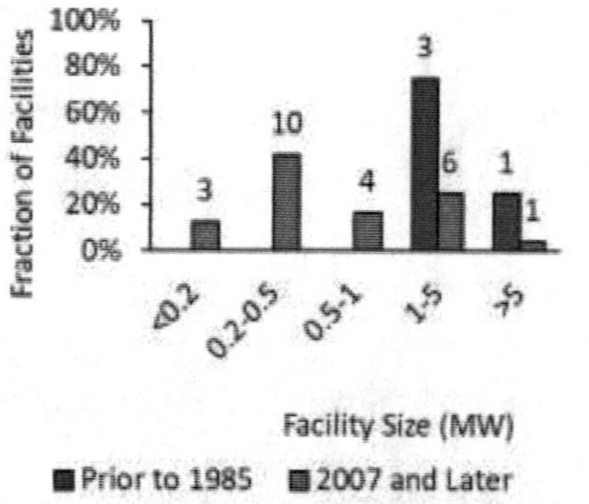

Figure 3. Small hydropower facility size by construction date. Larger facilities were built prior to 1985 than subsequently.

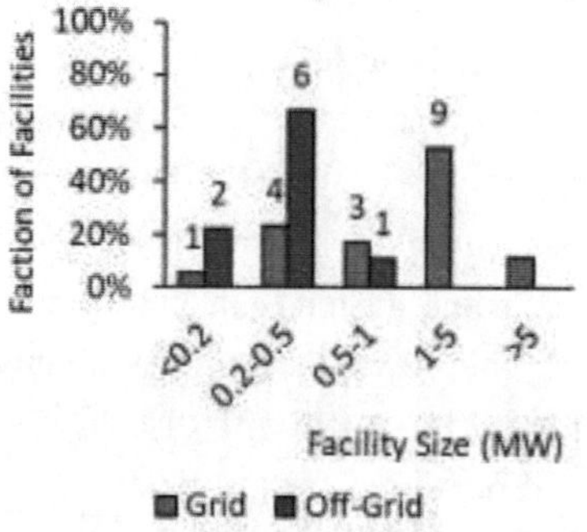

Figure 4. Grid connection status of small hydropower facilities. Larger facilities tend to be grid connected, while a substantial fraction of small facilities power off-grid systems.

Following policy changes to encourage development of generation sources, 24 facilities / 31 MW of capacity were commissioned since 2007. These facilities are smaller – averaging 1.3 MW/plant. A third of these new facilities are connected to off-grid power systems (Fig 4) in contrast to facilities built prior to 1985, which were all grid-connected. In addition, off-grid facilities tend to be smaller – averaging 0.2 MW – than on-grid facilities – averaging 1.9 MW.

2.2. Development potential

Based upon the survey data mentioned above, only a small fraction – 28 or 8% -- of potential small hydropower sites in Rwanda have been developed (Fig. 5). Therefore, a substantial number of possible development sites exist, for both on- and off-grid development. However, the survey data did not include an evaluation of potential cost or environmental conditions which might eliminate some sites from consideration. These could include sensitive watersheds, poor access to grid or population centers, or intermittent water flow.

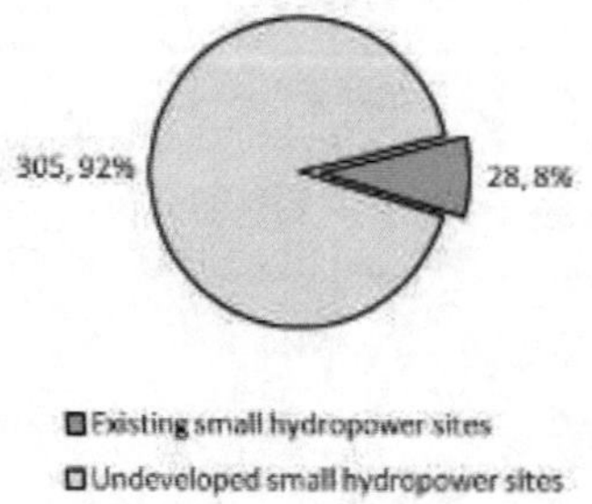

Figure 5. Comparison of developed small hydropower sites in Rwanda relative to a recent survey of locations with small hydropower potential.

2.3. Comparison of small hydropower and solar power development trends

All major Rwandan solar power plants are indicated in Table. 4. The table omits the many small off-grid systems, solar home systems or isolated PV-battery systems for schools, houses, clinics and other applications; no consolidated data was available for these systems.

Table 3. Small solar power plants in Rwanda.

Plant Name	Installed capacity (MW)	Year Commissioned	Connection
Mount Jali	0.25	2007	Grid connected
Rwamagana	8.5	2014	Grid connected
Nyamata	0.03	2016	Off-grid
Nasho	3.3	2015	Off-grid

Source[9, 10]

Comparing solar and hydropower development rate, both are developing at a high rate, but small hydropower is developing somewhat faster, due to a larger number of facilities and the relatively larger size of hydropower facilities.

3. Causes of rapid development in small hydropower

There has been rapid economic growth in sub-Saharan Africa, resulting in increased energy demand. This has pressured governments to explore all solutions to increase domestic energy supply[11].

3.1. Rwandan development incentives

Rwandan economy has performed well since 2000, with an average annual growth rate of nearly 8% [12]. The government has also identified several growth strategies, including development of the tourism and information and communications technology (ICT) sectors. Both sectors place a high priority on reliable and competitively priced energy, driving a substantial need to improve energy capacity and infrastructure [13]. Also, the Economic Development and Poverty Reduction Strategy II (EDPRS II) acknowledges that Rwanda's current energy system creates challenges for development. The plan articulates strategic areas of improvement in the energy sector. These include rapid expansion of energy supply capacity through grid and off-grid technologies [14].

To support development of energy sources, Rwanda has established laws meant to promote private sector involvement in energy development. The country has developed regulations including a renewable energy feed-in tariff for hydropower between 50 kW and 10 MW [15] and a simplified licensing method which exempts generation facilities of less than 50 kW from most licensing requirements [16].Additionally, the government has initiated a privatization of public services, including the main electricity utility. Benefits are provided to foreign investors, including tax holidays, import tariff reductions, assistance to access water and electricity, and assistance with obtaining visas and work permits [17]. Finally, the Rwanda Development Board (RDB) has launched a program to assist with business registration, licensing and obtaining tax incentives.

Stakeholder support promoting private sector involvement in energy sector investments has boosted power generation. An example is "Deutsche Gesellschaft für Internationale Zusammenarbeit (GIZ) in English German Corporation for International Cooperation" which promotes private investment in the energy sector through the Energising Development (EnDev) fund. Results-based financing was introduced into Rwanda by the UK Department for International Development (DFID) and implemented by EnDevin several countries, including Rwanda. The idea behind the approach is to reward companies for their previously-agreed and delivered results. For off-grid electrification, results-based financing provides companies incentives in form of subsidies for either in increasing sales of small electric systems (typically small PV systems) or in developing isolated village grids powered by renewable energy sources. Financing mechanisms are implemented by Urwego Opportunity Bank on behalf of EnDev Rwanda [18].

4. Opportunities

The combination of topology and hydrology make Rwanda an excellent location for small hydropower, while population density and other constraints likely would not encourage larger hydropower development that requires significant flooding of major valleys. As indicated earlier, many sites with power-producing potential remain undeveloped. The distributed nature of these sites could be complimentary to micro-grids, including off-grid power systems, as small hydropower facilities could be integrated into these facilities to provide 24/7 generation to compliment photovoltaic generation sources.

An upper bound on the potential for small hydropower can be estimated with several assumptions. First, of 333 identified sites, 28 have already been developed and we assume that 25% will prove unsuitable for environmental or other reasons. For example, some sites are likely to be in areas with sensitive plants or animals, or are in areas that are restricted from development, such as national parks or military installations. This leaves a total of 229 possible locations for development. If we further assume that new facilities will average 75% of the size of sites developed since 2007, potential generation capacity can be estimated as shown in Table 5. We assume 63%, or 143 facilities, would be grid-connected and average 1.5 MW, while 38% or 86 facilities, would be connected to off-grid systems and average 150 kW. Obviously, actual construction may deviate significantly from these assumptions, but they serve to scope the potential of small hydropower as guide to policy discussions. Given that Rwanda seeks to add approximately 300 MW of generation (from current 210 MW to 512 MW), potential generation from small hydropower sources could represent a non-trivial fraction of the solution – up to 60% of the desired increase in generation. Further, the distributed nature of the sites could support development of off-grid electrification, and thus not require extension of grid distribution systems.

Table 4. An estimate of potential generation possible from small hydropower in Rwanda.

	On Grid	Off-Grid	Total	Units
Possible Sites:	143	86	229	Sites
Approximate Size:	1.5	0.15	1.0	MW/site
Total Potential Gen:	209	13	222	MW

Producing energy domestically also provides a strategic advantage to Rwanda, as it does not require foreign currency reserves to pay for imported power in the case of electricity imports, or for fossil fuels, in the case of diesel-fueled generation.

Finally, the site survey [5] indicated that most small hydropower sites could be developed utilizing run-of-river methods. These methods do not require large dams to impound water – expensive and environmentally challenging construction projects.

5. Challenges

Small hydropower in Rwanda also faces several non-trivial challenges. First, the mountainous topology in Rwanda makes transmission and distribution relatively expensive. Mountain ranges up to 3,000m in height run across the country. Therefore, not all larger sites with multi-megawatt potential can be cost-effectively connected to Rwanda's transmission system, shown schematically in Fig. 6.

IOP Conf. Series: Earth and Environmental Science **133** (2018) 012013 doi:10.1088/1755-1315/133/1/012013

Few vendors provide equipment for small hydropower sites, and significant custom engineering is required for each facility. A 2012 study of hydropower in Colorado, USA, irrigation systems indicated that, while equipment existed, manufacturers for key components were few, were internationally-dispersed, and often required custom engineering for each installation [19, 20, 21]. As a result, installation engineering, and post-installation maintenance expertise and spare parts, present a challenge for Rwandan hydropower developers.

Rwanda is a landlocked country, increasing transportation costs for all imports, and particularly for heavy equipment. Since most hydropower plant components are manufactured outside of Africa, this increases the cost of hydropower plant investments. For example, of hydropower investment cost is estimated at 4000 US$/kW in Rwanda compared to 3829 US$/kW in Kenya [22, 23].

The integration and control of many small hydropower facilities into the national grid may present control challenges. Assuming a high penetration of small hydropower sites, the national grid control center may need automated mechanisms to dispatch small hydropower facilities. These mechanisms are not currently in place.

Finally, it is expected that climate change may reduce precipitation and flow Rwanda's from interconnect lakes, leading to a decline of electricity generation from small hydropower plants [24].

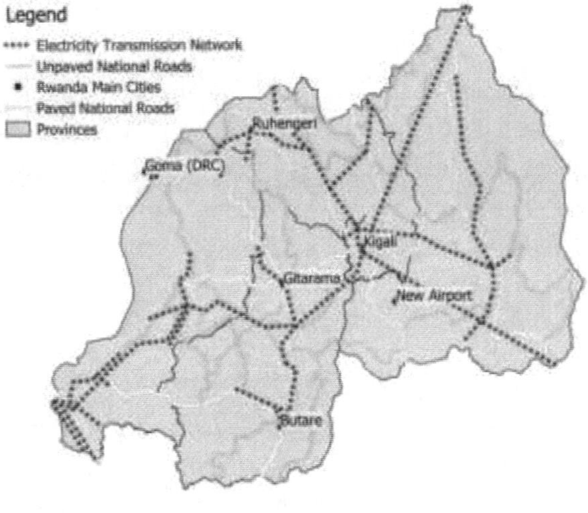

Figure 6. Approximate location of electricity transmission lines in Rwanda, 2012 data.

6. Conclusions and Policy Recommendations

Strong evidence indicates that small hydropower could contribute significant electricity production in Rwanda. The government actively encourages development, in contrast to many countries with more developed electricity systems. Many promising sites exist, distributed throughout the country. Considering the need for energy development and available hydropower resources, our analysis suggests several possible policy recommendations:

(1) Due to the difficulty of building energy transmission in Rwanda, it is advisable to encourage small hydropower use in off-grid systems;

(2) Investments should be made to develop the necessary support expertise for small hydropower technologies, including expertise for community-based micro-grids and training for plant maintenance to reduce operational cost.

(3) Small hydropower remains a niche business worldwide. Policies should be considered to attract investors willing to design and manufacture some hydropower plant components in Rwanda. This would allow solutions to be customized for Rwandan conditions, including:

a. Standardizing designs for run-of-river systems.
b. Improving designs to deal with high silt loads in Rwandan streams.
c. Providing local expertise, repair, and spare part service.

References

[1] Paish O 2002 Small hydro power: Technology and current status *Renew. Sustain. Energy Rev.***6**; 537–56.

[2] Rwanda Environment Management Authority 2008 *Sector Guidelines for Environmental Impact Assessment (EIA) for Hydro-power Development Projects in Rwanda.*

[3] Jaber J O 2012 Prospects and challenges of small hydropower development in Jordan *Jordan J. Mech. Ind. Eng.***6** 110–8.

[4] Ministry of Infrastructure 2015 *Energy Sector Strategic Plan.*

[5] Van der Plas R-J 2009 *Rwanda's Micro -Hydro Energy Market. Target Market Analysis.*

[6] *Energy overview by Rwanda Development Board (on line) http://www.rdb.rw/rdb/energy.html/accessed 3/11/2017.*

[7] Ministry of Infrastructure 2016 *Rural Electrification Strategy.*

[8] Renewable Energy Network for the 21st Century (REN21) 2016 *East-African renewable Energy and energy efficiency regional status.*

[9] *Solar energy overview by Rwanda Ministry of infrastructures* (On line) http://www.mininfra.gov.rw/index.php?id=85/ accessed on 7/9/2017

[10] Renewable Energy policy network for the 21st century 2016 *East African Community Renewable energy and Energy Efficiency Regional Status.*

[11] International Energy agency Africa Energy Outlook 2014 *focus on Energy Prospects in Sub-Saharan Africa.*

[12] United Nations Economic Commission for Africa 2015 *Rwanda profile.*

[13] Government of Rwanda 2011 *Rwanda National Export Strategy.*

[14] Minecofin 2013 *the Republic of Rwanda economic development and poverty reduction strategy.*

[15] Regulations No001/Energy/Rura/2012 of 09/02/2012 on Rwanda Renewable Energy feed in tariff.

[16] Regulation no 01/r/el-Ews/rura/2015 governing the simplified licensing framework for rural Electrification in Rwanda.

[17] Investment incetives by Rwanda development board *(on line) http://businessprocedures.rdb.rw/menu/25?l=en/ Accessed on 21/9/2017.*

[18] Improving energy in rural areas *(on line) https://www.giz.de/en/worldwide/28764.html/ Accessed on 22/3/2017.*

19] Rwanda Climate and geograph *(on line) http://www.aboutrwanda.com/rwanda-climate-and-geography/ Accessed on 6/10/2017.*

[20] B. Campbell and D. Zimmerle 2012, Evaluating Micro Hydropower Potential in Colorado's Irrigation Infrastructure, *Colorado Water.*

[21] B. Campbell 2012 Identification of spatial and topographical metrics for micro hydropower applications in irrigation infrastructure - *ProQuest, Colorado State University.*

[22] Claire Nelson, Rwanda energy sector overview *(on line): www.usaid.gov/powerafrica/rwanda. accessed 20/9/2017.*

[23] Pueyo A, Bawakyillenuo S and Osiolo H 2016 Cost and Returns on Renewable Energy in Sub-Saharan Africa: A comparison of Kenya and Ghana *Evid. Rep.*

[24] Rwanda Climate and geograph *(on line) http://www.aboutrwanda.com/rwanda-climate-and-geography/ Accessed on 6/8/2017.*

[25] Current electricity generation status *(on line) http://www.reg.rw/index.php/our-business/generation/465-current-electricity (accessed on 20/8/2017.*

[26] EDCL 2016 *Status of the Hydropower Sector in Rwanda.*

Appendix1

Table 5. Small hydropower plants currently operation in Rwanda.

S/N	Hydropower plant name	Commissioned Year	Installed capacity (MW)	Grid connected or off-grid
1	RUZIZI[1]	1957	3.5	Grid
2	GISENYI	1957	1.2	Grid
3	MUKUGWA	1982	10	Grid
4	GIHIRI	1984	1.8	Grid
5	NYAMYOTSI I	2007	0.1	Off-Grid
6	MUTOBO	2009	0.2	Off-Grid
7	AGATOBWE	2010	0.2	Off-Grid
8	MURUNDA	2010	0.1	Grid
9	RUKARARA I	2010	9	Grid
10	NYAMYOTSI II	2011	0.2	Off-Grid
11	RUSHUKI	2011	0.04	Off-Grid
12	KEYA	2011	2.2	Grid
13	NKORA	2011	0.68	Grid
14	KIMBILI	2011	0.3	Grid
15	RUGEZI	2011	2.2	Grid
16	NYABAHANGA	2012	0.2	Off-Grid
17	NSHILI	2012	0.4	Grid
18	MAZIMERU	2012	0.5	Grid
19	JANJA	2012	0.2	Off-Grid
20	GASHASHI	2013	0.2	Off-Grid
21	NYIRABUHOMBOH OMBO	2013	0.5	Off-Grid
22	MUKUNKWA II	2013	2.5	Grid
23	MUSARARA	2013	0.4	Grid
24	RUKARARA II	2014	2.2	Grid
25	GICIYE I	2014	4	Grid
26	GICIYE II	2016	4	Grid
27	GASEKE	2016	0.5	Grid
28	KIGASA	2016	0.2	Grid
	Total		47.5	

Source [25, 26]

IOP Conf. Series: Earth and Environmental Science **133** (2018) 012014 doi:10.1088/1755-1315/133/1/012014

A review of recent studies on the mechanisms and analysis methods of sub-synchronous oscillation in wind farms

Chenggen Wang[1], Qian Zhou[1], Shuning Gao[2], Jia Luo[2], Junchao Diao[2], Haoran Zhao[2] and Jing Bu[3]

[1] State Grid Jiangsu Electric Power Company Research Institute, Nanjing, China

[2] Shandong University, Jinan, China

[3] Nanjing University of Science and Technology, Nanjing, China

Abstract. This paper reviews the recent studies of Sub-Synchronous Oscillation(SSO) in wind farms. Mechanisms and analysis methods are the main concerns of this article. A classification method including new types of oscillation occurred between wind farms and HVDC systems and oscillation caused by Permanent Magnet Synchronous Generators(PMSG) is proposed. Characteristics of oscillation analysis techniques are summarized.

1. Introduction

Wind power generation has achieved a rapid development in the last decade, leading to a steady growth of wind power penetration. The existing transmission system need to be enhanced by using series compensation to accommodate the increased production by wind power. However, there is a possibility of an adverse interaction between the turbine generators and the series compensation, called the sub-synchronous resonance (SSR), which may result in damages to the generator shaft system.

So far, the SSR phenomenon and its mitigation techniques for conventional turbine-generator system have been well documented. However, the mechanism and mitigation techniques of sub-synchronous oscillation phenomenon occurred in series compensated wind farm as well as HVDC linked wind farm systems are still not fully understand. In recent years the emerging type of sub-synchronous phenomenon has been detected worldwide, for instance in the wind farm of USA (ERCOT) [1] and China (Qomul, Xinjiang; Guyuan, Hebei province) [2,3]. These events often accompanied with tripping of generator and damage of crowbar circuit which poses a severe threat to the stability of wind power transmission system. It becomes urgent to study on its mechanism and develop the effective damping algorithm.

The objective of this paper is to summarize the literatures in last few years about mechanisms and

Supported by the Science and Technology Project of State Grid Jiangsu Electric Power Company (J2017046)

analysis methods of sub-synchronous oscillation in modern wind power system.

2. Classification

Sub-Synchronous Oscillation (SSO) is a general concept. It defines an electric power condition where two power systems exchanges significant energy at one or more of the natural frequencies of the combined system below the synchronous frequency. It is also called Sub-Synchronous Interaction (SSI) in number of studies [4,5]. According to to the different elements involved, the phenomenon can be further distinguished and classified into Sub-Synchronous Resonance, Sub-Synchronous Torsional interaction(SSTI) and Sub-Synchronous Control Interaction (SSCI) [6,7].

2.1. Sub-Synchronous Resonance

SSR is a condition where a series capacitor compensated system exchanges significant energy with a turbine-generator at a frequency below the synchronous frequency. SSR can be further classified into three types: induction generator effect (IGE), torsional interaction (TI) and torque amplification (TA).

IGE is a purely electrical phenomenon and takes place in the power system, which requires a high degree of series compensation [2,8,9]. Different from IGE, TI is not only an electrical phenomenon but also involves mechanical dynamics. It generally occurs when the natural frequency of generator shaft is close to the frequency of induced sub-synchronous electromagnetic torque. However, as the shaft stiffness of wind turbine is rather low, TI is rarely induced. Recent studies clarified that IGE instead of TI is the major cause of SSR instability in the case of variable speed-DFIG based wind generation [9].

TA is also referred to as "transient torque" or "transient SSR" occurs due to faults in the network or during switching operations. Although TA can cause severe mechanical torsional oscillations in shaft system, there is no such record of severe shaft damage due to TA occurred in wind farm.

SSR is proved to be possible in DFIG-based wind farm interconnected with series compensated networks. SSR occurred in wind farm is mainly affected by three factors 1) series compensation level; 2) wind speed; and 3) current loop gains [2,8-11]. Ref. [12] assessed SSR under uncertain wind speed conditions by using a piecewise probabilistic collocation method. The location of DFIGs is considered as a factor of SSR in ref. [11], which confirmed that DFIGs closer to PCC is more seriously affected by series compensation degree.

2.2. Sub-Synchronous (Control) Interaction

There are narrow sense and broad sense for the concept “SSCI”, the narrow sense SSCI defines an interaction between series compensated transmission system and a power electronics control system (such as HVDC link, SVC or wind turbine control system) [6,13].The concept of narrow sense of SSCI is often confused with the concept of IGE [14] since it has been widely proved that the characteristics of IGE between wind farm and series compensation is strongly affected by proportional gains in current control loops of DFIG converters [8,9,13,15,16]. It has also confirmed that the DFIG input measuring filters have significant impact on SSCI [17].

There are some new types of SSR/SSO appeared recently which is similar to SSCI but difficult to incorporate them into existing categories.

Field experience has shown that the SSO can occur between the wind farms and HVDC, since the collection bus power of transmission system is often connected to a weak system [18-21]. The oscillation is arguably originated by the interaction between the wind energy conversion system inverter controller and the HVDC rectifier [18,21]. Full converter of wind turbine generator divides the system into two such that the oscillatory modes have participation only from the states located at one of the two sides. Hence, the generator, turbine and machine side controller models are not included for the modes between wind farms and HVDC. Possible origins of the SSO and resonances observed in the interconnection of a WF and a VSC-based HVDC system is discussed in [18,21].

Recent studies also shown that Permanent Magnet Synchronous Generators(PMSG) would be exposed to unstable SSO [22]. The frequency and stability of such SSO depends on many factors including the short circuit ratio and the control strategies & parameters of wind turbine generator converters [5]. Ref. [22] confirmed that the delay time of feed forward voltage link is the direct drive to SSO in PMSGs connected with power grid. Oscillatory modes present in an PMSGs based offshore wind farm system connected to a VSC-HVDC system using modal analysis tools is proposed in [23], the characteristics of the modes are influenced by the PMSG parameters, such as GSC controller, PLLs, LCL filter, and controller parameters of VSC-HVDC.

To incorporate these new types of SSR/SSO into existing categories, it is necessary to reconsider SSR/SSO definitions and patterns. A concept of broad sense of SS(C)I is proposed in [24]. SS(C)I defines an coupling interaction between two power electronics control systems or between the electronics control system and the grid-side elements. The inherent oscillatory mode of such interaction is hard to find from the machine or grid side.

2.3. Sub-Synchronous Torsional Interaction

SSTI defines the interaction of a turbine-generator near a power electronic controller when the mechanical system resonates with the negative damping of the controller at sub-synchronous frequencies. SSTI is different from SSR because there is no system resonance involved. Due to the low-shaft stiffness coefficient, SSTI rarely happened in wind turbine generator system [2]. Ref. [20] investigated SSTI associated with a practical multi-terminal MMC-based VSC-HVDC.

3. Oscillation analysis techniques

The most common techniques available for the study of sub-synchronous oscillation phenomenon can be classified into three types: impedance based analysis, eigenvalue analysis and time domain simulation analysis. Time domain simulation analysis implies the utilization of simulation program such as PSCAD/MATLAB/DigSILENT, which is widely used to validate results obtained by impedance based analysis and eigenvalue analysis [18,25,26]. The accuracies of various analytical tools used for SSCI problem identification is compared in ref. [17].

3.1. Impedance based analysis

According to the difference of stability criterion, impedance based analysis can be further classified into two types: impedance based frequency scanning method and impedance based Nyquist stability analysis.

The frequency scanning method is a basic technique for analysing SSR, whish is widely used in industry. In this technique, the equivalent resistance and reactance of the system as viewed from driving point are computed as a function of frequency. Aspects such as the injection/measurement point or the choice of injected signal waveform need to be considered in frequency scanning method are discussed in detail [25]. The SSR may induced once the reactance is zero (or close to zero), and resistance is negative at sub-synchronous frequency. Another potential risk assessment for SSCI issue is presented [27].

Impedance based Nyquist stability analysis is a well-established technique to analyse such interconnected systems. The essence of this method is to partition the system under study into a source subsystem and load subsystem. The ratio of the equivalent source output impedance to the equivalent load input impedance must satisfy the Nyquist stability criterion to guarantee the stability of interconnected source-load system [28]. Equivalent impedance models of a DFIG along with its rotor-side converter (RSC) and grid-side converter is derived in ref. [8]. The improved impedance model which incorporates DFIG's full-scale control system is proposed in [29]. Another refined approach that represents both the wind farm and the grid using two-by-two impedance matrices is proposed in [4].

Nyquist criterion is also widely used in recent publications for the analysis of sub-synchronous oscillations between wind farms and HVDC power transmission system [18,21,30].

A detailed model of wind farm and HVDC impedance model were derived and the potential resonance points were identified by using this criterion [18]. A reformulated positive-net-damping criterion based on Nyquist stability analysis to evaluating net damping between HVDC and wind farm is proposed in [21].

3.2. Eigenvalue analysis

Eigenvalue analysis is the most commonly used analytical method, based on mathematical model of system under study using a set of linear ordinary differential equations. The standard form of differential equations can be written as following:

$$\dot{x} = Ax + Bu \tag{1}$$

The eigenvalues are then defined as the solution to equation:

$$det[\lambda U - A] = 0 \tag{2}$$

Where λ is eigenvalue vector of the differential equation.

The eigenvalue can clearly indicate the frequencies and damping of oscillation modes, i.e. the relationship between the system parameters (series compensation level, PI parameters of current controller, wind speed, number of wind turbine generators) and oscillation characteristics can be easily shown by the movement of eigenvalues [6,9]. D.H.R. Suriyaarachchi et al. provide a two steps procedure to study sub-synchronous interactions, a frequency scanning method is applied in step.1 and a detailed small signal eigenvalue analysis is performed in step.2 to obtain information about oscillatory modes [15]. Ref. [23] used participation factor analysis and confirmed that characteristics of the modes are influenced by the PMSG parameters, such as GSC controller, PLLs, LCL filter, and controller parameters of VSC-HVDC.

However, the eigenvalue analysis approach requires a detailed representation of both the mechanical and electrical systems. Thus, higher order models than the 2nd or 4th order models are therefore required. The model parameters for these high order models are not readily known [25].

Table 1. Overview of research publications on sub-synchronous oscillation in wind farm.

SSR	SSCI	SSCI(PSMG)	SSCI(HVDC)	SSTI
[2, 11-13, 15, 16, 31, 32]	[4, 6, 17, 27, 29, 33]	[5, 22, 34]	[18, 21, 23, 26]	[20]

4. Conclusion

This paper is a summary of recent studies about SSI/SSO observed in wind field and the wind power transmission system. The conclusions can be summarized as follow: SSR, SSCI and SSTI are three main classifications of subsynchronous phenomenon. SSTI caused by controlling unit in the components of the WF and the HVdc system is an emerging oscillation type been widely observed and studied recently. The frequency scanning analysis, eigenvalue analysis, time domain simulation analysis and impedance based Nyquist stability analysis are the most commonly used analyzing method for the studies of the SSI/SSO.

References

[1] J Adams, C Carter, and S H. Huang 2012 ERCOT experience with Sub-synchronous Control Interaction and proposed remediation," Proceedings of the IEEE Power Engineering Society Transmission and Distribution Conference 1-5

[2] X Xie, X Zhang, H Liu, H Liu, Y Li and C Zhang 2017 Characteristic Analysis of Subsynchronous Resonance in Practical Wind Farms Connected to Series-compensated Transmissions *J.IEEE Transactions on Energy Conversion*.vol. C **8969**

[3] M Li, Z Yu, T Xu, J He, C Wang, X Xie and C Liu 2017 Study of complex oscillation caused by renewable energy integration and its solution *J.Dianwang Jishu/Power System Technology*. 4 **41** 1035–42

[4] W Ren and E Larsen 2017 A Refined Frequency Scan Approach to Sub-Synchronous Control Interaction (SSCI) Study of Wind Farms *J. IEEE Transactions on Power Systems*. 5 **31** 3904 –12

[5] H Liu, X Xie, J He, T Xu, Z Yu, C Wang and C Zhang 2017 Subsynchronous Interaction between Direct-Drive PMSG Based Wind Farms and Weak AC Networks *J.IEEE Transactions on Power Systems*. C **8950**

[6] A E Leon and J A Solsona 2015 Sub-synchronous interaction damping control for DFIG wind turbines *J.IEEE Transactions on Power Systems*. 1 **30** 419–28

[7] V B Virulkar and G V Gotmare 2016 Sub-synchronous resonance in series compensated wind farm:A review *J.Renewable and Sustainable Energy Reviews*. **55** 1010–29

[8] Z Miao 2012 Impedance-model-based SSR analysis for type 3 wind generator and series-compensated network *J.IEEE Transactions on Energy Conversion*. 4 **27** 984–91

[9] L Fan, R Kavasseri, Z L Miao and C Zhu 2010 Modeling of DFIG-based Wind farms for SSR analysis *J.IEEE Trans. Power Delivery*. 4 **25** 2073–82

[10] H Liu, X Xie, C Zhang, Y Li, H Liu and Y Hu 2017 Quantitative SSR Analysis of Series-Compensated DFIG-Based Wind Farms Using Aggregated RLC Circuit Model *J.IEEE Transactions on Power Systems*.1 **32** 474–83

[11] X. Zhu and Z. Pan 2017 Impedance-Model-Based SSR Study considering DFIGs at Different

Locations 5-9

[12] W Chen, X Xie, D Wang, H Liu and H Liu 2017 Probabilistic Stability Analysis of Subsynchronous Resonance for Series- Compensated DFIG-based Wind Farms *J.IEEE Transactions on Sustainable Energy*. C **3029**

[13] H Liu, X Xie, Y Li, H Liu and Y Hu 2017 Mitigation of SSR by embedding subsynchronous notch filters into DFIG converter controllers *J.IET Generation, Transmission & Distribution*. 11 **11** 2888–96

[14] ABB Technical Memorandum. Comment on ERCOT NPRR562 and Other Informational Terms [15] H A Mohammadpour, E Santi and A Ghaderi 2014 A Procedure to Study Sub-Synchronous Interactions in Wind Integrated Power Systems *J.IET Generation, Transmission & Distribution*. 12 **8** 1998–2011

[16] M Ghafouri, U Karaagac, H Karimi, S Jensen, J Mahseredjian and S. O. Faried 2017 An LQR Controller for Damping of Subsynchronous Interaction in DFIG-Based Wind Farms *J.IEEE Transactions on Power Systems*. C **8950**

[17] U Karaagac, J Mahseredjian, S Jensen, R Gagnon, M Fecteau and I Kocar 2017 Safe Operation of DFIG based Wind Parks in Series Compensated Systems *J.IEEE Transactions on Power Delivery*. C **8977**

[18] M Amin and M Molinas 2017 Understanding the origin of oscillatory phenomena observed between wind farms and hvdc systems *J.IEEE Journal of Emerging and Selected Topics in Power Electronics*. **5** 378–92

[19] J Lv, P Dong, G Shi, X Cai, H Rao, and J Chen 2014 Subsynchronous oscillation of large DFIG-based wind farms integration through MMC-based HVDC *International Conference on Power System Technology: Towards Green, Efficient and Smart Power System*. 2401-2408

[20] D Sun, X Xie, Y Liu, K Wang and M Ye 2016 Investigation of SSTI Between Practical MMC-based VSC-HVDC and Adjacent Turbo generators through Modal Signal Injection Test *J.IEEE Transactions on Power Delivery* **32** 2432-41

[21] M Cheah-Mane, L Sainz, J Liang, N Jenkins, and C E Ugalde Loo 2017 Criterion for the Electrical Resonance Stability of Onshore Wind Power Plants Connected through HVDC Links *J.IEEE Transactions on Power Systems*. **32** 4579-4589

[22] H Qifeil, H Zhiguol, and Z Baohui 2016 The Research of Sub Synchronous Oscillation in PMSG Wind Farm *IEEE PES Asia-Pacific Power and Energy Conference*

[23] L P Kunjumuhammed, B C Pal, R Gupta, and K J Dyke 2017 Stability Analysis of a PMSG-Based Large Onshore Wind Farm Connected to a VSC-HVDC *J.IEEE Transactions on Energy Conversion* **32** 1166-1176

[24] X Xie, L Wang, J He, H Liu, C Wang, and Y Zhan 2017 Analysis of subsynchronous resonance/oscillation types in power systems *J.Dianwang Jishu/Power System Technology*, **41** 1043-1049

[25] B Badrzadeh, M Sahni, Y Zhou, D Muthumuni and A Gole 2013 General Methodology for Analysis of Sub-Synchronous Interaction in Wind Power Plants *J.IEEE Transactions on Power Systems*. **28** 2 1858-1869

[26] J Lyu, X Cai, and M Molinas 2017 Optimal Design of Controller Parameters for Improving the

IOP Conf. Series: Earth and Environmental Science **133** (2018) 012014 doi:10.1088/1755-1315/133/1/012014

Stability of MMC-HVDC for Wind Farm Integration *J.IEEE Journal of Emerging and Selected Topics in Power Electronics* C **6777** 1-1

[27] M A Chowdhury, M A Mahmud, W Shen and H R Pota 2017 Nonlinear Controller Design for Series Compensated DFIG-Based Wind Farms to Mitigate Subsynchronous Control Interaction *IEEE Transactions on Energy Conversion*. **32** 707-19

[28] J Sun 2011 Impedance-Based Stability Criterion for Grid-Connected Inverters *J.IEEE Transactions on Power Electronics* **26** 11 1-10

[29] H Liu, X Xie, C Zhang, Y Li, H Liu and Y Hu 2017 Quantitative SSR Analysis of Series-Compensated DFIG-Based Wind Farms Using Aggregated RLC Circuit Model *J.IEEE Transactions on Power Systems* **32** 474-483,

[30] Han chao Liu and Jian Sun 2014 Voltage Stability and Control of Onshore Wind Farms With AC Collection and HVDC Transmission *J.IEEE Journal of Emerging and Selected Topics in Power Electronics* 4 **2** 1181-89

[31] L B Shi, J L Su and L Z Yao 2017 SS resonance analysis of complex power system incorporating wind power J.*IET Renewable Power Generation*. 3 **11** 305-312

[32] M A Chowdhury, G M Shaullah and S Member 2017 SSR Mitigation of Series-Compensated *J.IEEE Transactions on Power Systems* C **8950**

[33] M T Ali, M Ghandhari and L Harnefors 2017 Mitigation of Sub-Synchronous Control Interaction in DFIGs using a Power Oscillation Damper *PowerTech*

[34] T Bi, J Li, P Zhang, E Mitchell-Colgan and S Xiao 2017 Study on response characteristics of grid-side converter controller of PMSG to sub-synchronous frequency component *J.Renewable Power Generation*. 7 966-72

Optimization of power rationing order based on fuzzy evaluation model

Siyuan Zhang, Li Liu, Peiyuan Xie, Jihong Tang and Canlin Wang
State Grid Hunan Electric Power Company, China

Abstract. With the development of production and economic growth, China's electricity load has experienced a significant increase. Over the years, in order to alleviate the contradiction of power shortage, a series of policies and measures to speed up electric power construction have been made in china, which promotes the rapid development of the power industry and the power construction has made great achievements. For China, after large-scale power facilities, power grid long-term power shortage situation has been improved to some extent, but in a certain period of time, the power development still exists uneven development. On the whole, it is still in the state of insufficient power, and the situation of power restriction is still severe in some areas, so it is necessary to study on the power rationing.

1. Introduction

With the rapid development of national economy and the continuous improvement of people's living standard, the peak valley difference of power grid is increasing day by day. As the power in the current level of technology cannot be a lot of storage, and the power system of the hair, lose, power distribution needs to be carried out at the same time[1].Once there is any problem in the power system, it will spread to other links, and even cause irreparable damage. Therefore, switching power limit during overload of power grid has a great significance of ensuring the safe operation of power grid, guaranteeing the quality of power consumption and alleviating the peak power shortage. Making reasonable power rationing optimization scheme can not only increase the income of enterprises, but also optimize the allocation of resources, so as to realize the sustainable development of the country.

2. Analysis of influencing factors on optimization of power rationing order

Power rationing refers that on the basis of satisfying the constraint condition of the generator set or the power system operation, when the power grid is overloaded, the part of the load is cut off by the power supply enterprise, through the power cut load to ensure that the time to participate in the operation of the unit and the unit in the run-time output are under normal circumstances, so that the total operating costs of the system and the loss of social benefits at least. The power rationing is of great significance to improve economy and reliability of power system operation and the utilization efficiency and comprehensive utilization level of the power system.

Many factors affect the power rationing order optimization scheme determined, these factors are uncertain, and mutually influence each other, which can be divided into economic, technological, social and environmental factors. Economic factors mainly include potential value of power customers, economic value of power customers, power supply interruption loss and other factors[2];technical factors include power market equilibrium, power market operation, power system trends and other factors; social factors include household electricity consumption, political electricity conservation, high risk customer power supply and other factors; environmental factors include social sustainable development, ecological environment and other factors. Specific factors are shown in the following figure 1.

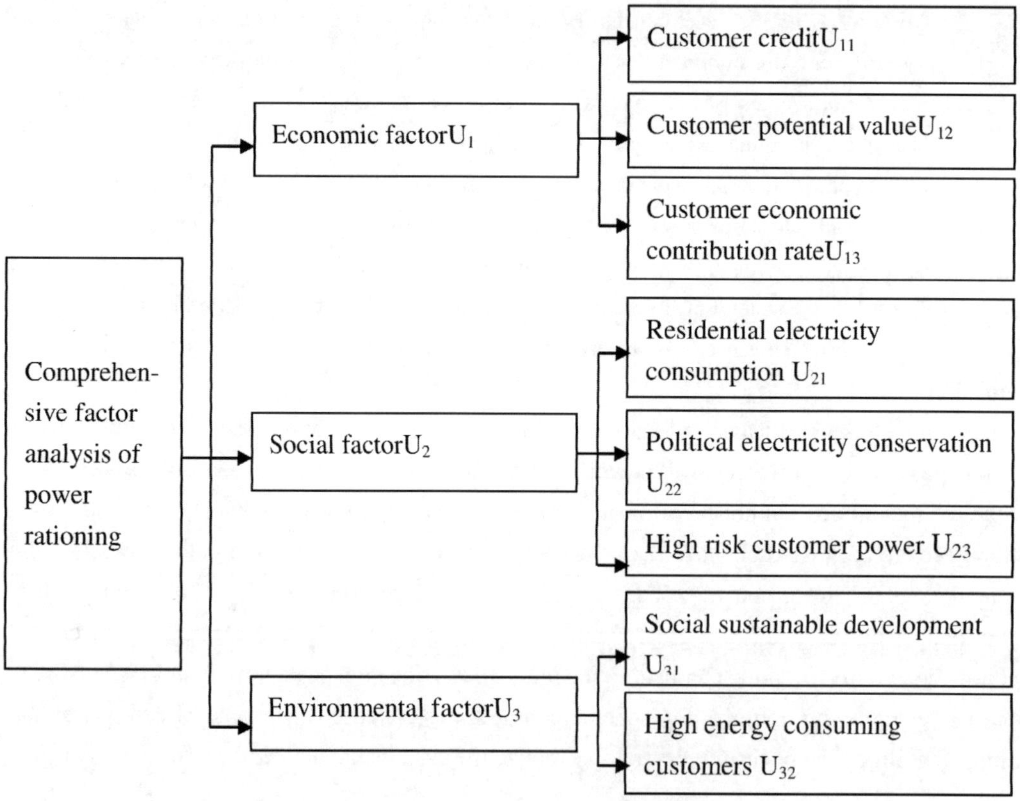

Figure 1. Hierarchical relationship of factors influencing power rationing

3. Evaluation model of power rationing order optimization

There are many factors or indexes involved in the power rationing, which requires the comprehensive evaluation of the things according to these factors or multiple indexes. So you can't judge things only from one factor. Fuzzy comprehensive evaluation decision is a very effective multi factor method for comprehensive evaluation of things affected by many factors [3].In this paper, the fuzzy evaluation model is used to study the comprehensive factors influencing the power rationing of the power system, and the evaluation model of the power supply system is established, which provide us an evidence for developing a reasonable policy decision of power rationing order optimization.

3.1. Fuzzy comprehensive evaluation method

Fuzzy comprehensive evaluation refers to finding a weight vector A={a1,a2,…,ai,…,am}∈F(U), and a fuzzy transformation f from U to V, that is, to make a judgment of each factor ui individually. f(ui)=(ri1,ri2,…,rin)∈F(V), i=1,2,…,m, in view of the above, the fuzzy matrix is formed, that is R=[rij]∈F(U×V),among them,rij represent factor ui, which have the degree of comment vj.That is to say, the membership degree of vj in fuzzy set B.

Thus it can be seen that the mathematical model of fuzzy comprehensive evaluation involves three factors:

(1)Factor set $U=\{u_1, u_2, u_3, \dots, u_m\}$;

(2)Decision set $V=\{v_1, v_2, v_3, \dots, v_n\}$;

(3)Single factor judgment f: $U\rightarrow F(V), u_i| \rightarrow f(u_i) = (r_{i1}, r_{i2}, \dots, r_{in}) \in F(V)$.

F induced fuzzy relation:$R_f \in F(U \times V)$, among them, $R_f(u_i, v_j) = f(u_i)(v_j)$, so the R_f can form fuzzy matrix.

$$R=\begin{pmatrix} r_{11} & \cdots & r_{1n} \\ \vdots & \ddots & \vdots \\ r_{m1} & \cdots & r_{mn} \end{pmatrix} \quad (1)$$

$A=\{a_1,a_2,\dots,a_i,\dots,a_m\}$ is the weight fuzzy vector on factor set U, the fuzzy set R=A× R on decision set V is transformed by R transform, so (U,V,R) constitutes a comprehensive evaluation model.

3.2. Basic steps of power rationing order optimization evaluation model

In the index system of power rationing, due to the type of each index set carrying different information, each index subsystem and specific index items have different degrees of action in the description process. Therefore, the comprehensive index value of power rationing is a weighted summation relation, that is:

$$S = \sum w_i f_i(I_i) \; (i = 1,2, \dots, n) \quad (2)$$

Where:$f_i(I_i)$is index measurement value of the index Ii; wi is the weight coefficient of each index, meeting $\sum w_i = 1, w_i \in [0,1]$.$f_i(I_i)$ can be obtained through the collection, statistics, analysis and comparison of the original data.The weight value of the comprehensive index system is obtained through the model.

3.2.1. Determine the level of factors and levels of factors, establish evaluation index set

In this paper, the factors that influence the power rationing factor are divided into economic factors, social effects and environmental impacts. Systematic analysis is made on the power rationing influencing factors, and the factors are classified according to the hierarchy, and the hierarchical relation is established as shown in figure 1. According to the order picture we can establish the following power rationing index system.

First evaluation index set is U.

IOP Conf. Series: Earth and Environmental Science **133** (2018) 012015 doi:10.1088/1755-1315/133/1/012015

U = {U1, U2… Um} (m is the number of first indicators).

Second evaluation index set is Ui.

Ui={Ui1,Ui2,…Uin} (i=1,2,…,m) (n is the number of second evaluation indicators below first indicators).

3.2.2. Determine the weight set of each index of evaluation index

The commonly used methods of determining weights are Delphy Fa, AHP [4], factor analysis, and entropy method and so on. In these methods, each method has its own advantages and disadvantages. According to the characteristics of power limits, this has different influence on different users. First, we use the method of expert consultation to construct the two comparison matrix, and then use the root method or the sum product to find the eigenvector and eigenvalue of the matrix, and carry out the consistency test to get the weight of each index.

Set the weight set of the first level index: A={$a_1, a_2, \ldots, a_i, \ldots, a_m$}, ($0 \le a_i \le 1, \sum a_i = 1 (i = 1,2,\ldots,m)$);

Set the weight set of the second level index: A={$a_{i1}, a_{i2}, \ldots, a_{ij}, \ldots, a_{in}$}, ($0 \le a_{ij} \le 1, \sum a_{ij} = 1 (j = 1,2,\ldots,n)$);

3.2.3. Determine the rating scale of evaluation pointers

According to the requirements of power supply reliability and the degree of loss or influence caused by interruption of power supply in politics and economy, the evaluation of each factor can be divided into four grades, the standard set of degree evaluation can be expressed as: V={V1,V2,V3,V4}, the corresponding importance can be described as {lower, low, medium, high}, they are assigned to (4,3,2,1) respectively. If the pointer level is between two adjacent grades, the corresponding score is 3.5, 2.5, 1.5, and 0.5.

3.2.4. Score and construct evaluation matrix

Combined with the evaluation of r experts on the evaluation data, which can be noted as d_{ijk}(i=1,2,…,m; j=1,2,…,n; k=1,2,…,r). From above, we can get the evaluation matrix D:

$$D=\begin{pmatrix} d_{111} & \cdots & d_{11r} \\ \vdots & \ddots & \vdots \\ d_{mn1} & \cdots & d_{mnr} \end{pmatrix} \quad (i=1,2,\ldots,m;\ j=1,2,\ldots,n;\ k=1,2,\ldots,r) \tag{3}$$

3.2.5. The grey number is determined, calculated and evaluated

In order to truly reflect the degree of belonging to a certain class, it is necessary to determine the evaluation grey class which is to determine the grey grade number, grey number and grey number of whitening weight function. According to the evaluation grade, the evaluation grey class is determined. Calculate the grey number according to the method of empowerment based on function driven. Make a comprehensive evaluation of the evaluation index Uij, the comprehensive evaluation result is denoted as Bi. Then make a comprehensive evaluation of the first level index Ui, the comprehensive evaluation result is denoted as B.

3.2.6. Calculate the comprehensive evaluation value and evaluate the order

Finally calculate the comprehensive evaluation value according to the above evaluation grey class value vector; we can get the evaluation order that can provide guidance for us.

In the formulation of optimization scheme, for the evaluation of the importance of multiple power users, it can be sorted and analysed by comparing the influence degree of the size of comprehensive evaluation value, which is measured from the benefits of power supply enterprises and the impact on society. For the evaluation of electric power load, through comprehensive evaluation values determine the degree of importance. It can also evaluate the importance of value factors and focus on controlling customers with large load and little influence.

4. Conclusion

Influence factors of blackouts involve multiple factors or indicators, on the basis of the above analysis, a set of analysis and index system of power rationing is established from four aspects: economic factors, technical factors, social factors and environmental factors, which affect the power system power rationing. Through the evaluation model, a comprehensive qualitative evaluation on the factors that affect the power rationing of power system is made. And it provides a basis for the power supply enterprises to develop a reasonable optimization scheme for power rationing.

5. References

[1] Yao Jiangang, Zhang Jian, Yin Chelai. Power market operation and software development. *Beijing: China Electric Power Press*,2002

[2] S. Ashok, Rangan Banerjee. An Optimization Mode for Industrial Load Management. *IEEE Trans on Power Systems*, 2001, **16(4)**: 879~884

[3] Wang Wenqing.Study on Optimization of power order switching off the power supply enterprise [D].*North China Electric Power University (Beijing)*, 2009

[4] Mu Yongzheng, Lu Zongxiang, Qiao Ying, Wang Yang, Huang Han, Zhou Qinyong, Han Jiahui. Comprehensive evaluation index system of power grid security and benefit based on fuzzy evaluation of multi operator hierarchy [J]. *Power Grid Technology*, 2015, **39 (01)**: 23-28.

[5] Mingming Zhang, Junquan Wang, Bo Peng, Hao Wu. An orderly power use decision making method with multi time scale coordination [J]. *Automation of Electric Power Systems*, 2015, **39 (01)**: 23-28.

[6] Jun Chen. Risk assessment of regional power grid operation based on maintenance mode [D]. South China University of Technology, 2013.

[7] CHEN Guang-xian,YANG Xu-guang. Analysis and recommendations on orderly power utility in regional power grid [J]. Power Demand Side Management,2015,**(5)**:42-43,60.

[8] YU Bing,LEI Jun-zhao. Understunding of Orderly Power Consumption[J]. Coal Technology,2012,**(12)**:278-279.

[9] ZHANG Xueli,DONG Xueping.Algorithm of orderly power consumption managementbased on non-cooperative game [J]. Journal of Hefei University of Technology(Natural Science),2017,(7):922-925.

[10] ZHANG Jiali, YI Pan. Research on the problem of ordered electricity use [C]. China Institute of electrical engineering, electric power system automation, Specialized Committee 2012 academic exchange.2012:1-7.

[11] MIAO WANG, XI-ZHAO WANG.A RESEARCH ON WEIGHT ACQUISITION OF WEIGHTED FUZZY PRODUCTION RULES BASED ON GENETIC ALGORITHM[C].2006 International Conference on Machine Learning and Cybernetics.2006:2208-2211.

IOP Conf. Series: Earth and Environmental Science **133** (2018) 012016 doi:10.1088/1755-1315/133/1/012016

Review of Recent Development of Dynamic Wind Farm Equivalent Models Based on Big Data Mining

Chenggen Wang[1], Qian Zhou[1], Mingzhe Han[2], Zhan'ao Lv[2], Xiao Hou[2], Haoran Zhao[2] and Jing Bu[3]

[1]State Grid Jiangsu Electric Power Company Research Institute, Nanjing, China

[2]Shandong University, Jinan, China

[3]Nanjing University of Science and Technology, Nanjing, China

Abstract. Recently, the big data mining method has been applied in dynamic wind farm equivalent modeling. In this paper, its recent development with present research both domestic and overseas is reviewed. Firstly, the studies of wind speed prediction, equivalence and its distribution in the wind farm are concluded. Secondly, two typical approaches used in the big data mining method is introduced, respectively. For single wind turbine equivalent modeling, it focuses on how to choose and identify equivalent parameters. For multiple wind turbine equivalent modeling, the following three aspects are concentrated, i.e. aggregation of different wind turbine clusters, the parameters in the same cluster, and equivalence of collector system. Thirdly, an outlook on the development of dynamic wind farm equivalent models in the future is discussed.

1. Introduction

The penetration of wind energy in power system has been growing due to its interminable and mild environment effects. In 2016, more than 54GW of wind power was installed across the global market. The cumulative capacity has reached to 486.8GW with a capacity growth of 12.6% [1]. Nowadays, wind power has become an important part of grid power supply. In the meantime, the large-scale wind power integration has brought a lot of stability problems, which endanger the operation security of power systems. Therefore, dynamic modeling of wind power is highly required for power system analysis.

The dynamic model of wind farm includes detailed model and equivalent model. With the growing size of the modern wind farms, detailed model is no longer applicable due to its complexity and long simulation time [2][3]. In order to reduce the computational burden and the simulation time, equivalent modeling method is required and developed. In the specific studies, reasonable equivalence simplification does not affect the accuracy of the analytical results.

Supported by the Science and Technology Project of State Grid Jiangsu Electric Power Company (J2017046)

In recent years, big data mining has been developing rapidly. It is gradually realized that the deeper and more important information hidden behind the data can describe the overall characteristic of the data and predict the development trend. These information can play an important role in the decision-making process [4]. Thus, some researchers began to apply big data mining to the dynamic equivalent modeling of wind farms.

This paper mainly reviews and analyzes the recent development of dynamic equivalent modeling of wind farms based on the big data mining method, which consists of the following three parts. The first part is the wind speed of the wind farm, including the wind speed prediction, the wind speed distribution of wind farms, and the equivalence of wind speed. The second part is the dynamic single wind turbine equivalent modeling of wind farm, including the selection of equivalent parameters and the parameter identification. The third part is the dynamic multiple wind turbine equivalent modeling of wind farms, which further includes three aspects: clustering of wind farms, parameter identification of clustered wind turbines, and equivalence of wind farm collector systems.

2. Wind speed in the wind farm

For dynamic wind farm equivalent modeling and model verification, the relevant data regarding wind speed and wind direction is needed. Considering factors such as wake effect, turbine distribution etc., the wind speed distribution and the incoming wind speed of each wind turbine could be calculated. Similarly, the equivalent wind speed of every equivalent wind turbine could also be calculated.

2.1. The wind speed prediction in the wind farm

To get the natural wind speed and direction of the upstream of the wind farm, neural network-based models [5] and Weibull distribution models are widely used.

2.2. The wind speed distribution in the wind farm

The distance between wind turbines is limited due to collector system and geographical constraint, especially in the large-scale wind farms. Wind turbines extract energy from the wind to produce electricity. Accordingly, the wind downstream of the turbine has a lower energy than the wind upstream of the turbine. This shadowing effect from upstream turbines on other turbines further downstream is referred to as the wake effect [6].

Based on the wind attenuation factor, a method to calculate the wakes of multiple turbines is proposed in [5].

2.3. The application of equivalent wind in the wind farm

Once the data of natural wind in the upstream of the wind farm is calculated, the equivalent power curve could be obtained. The method to calculate the equivalent wind of the aggregated turbines such as average wind speed method and the method based on typical equivalent power curve is widely used. Based on the neural network static model, dynamic real-time simulation model was developed in [7].

Two conventional approaches for the equivalent steady state modeling of the wind farm are discussed: a) aggregate-measured power curve, obtained as an averaged aggregate representation of the outputs from all wind turbines, and b) cluster-measured power curve, obtained using the support vector clustering technique [8].

3. Dynamic single turbine equivalent models of the wind farm based on big data mining

The use of single turbine equivalent model to represent the dynamics of large-scale wind farm can significantly simplify the calculating process during simulation, comparing to the accurate detail model. The following two steps are used to build an equivalent single turbine model: choosing equivalent parameters and calculating parameters.

3.1. Choose equivalent parameters

To simplify the wind turbine model, it is the first step to choose typical parameters that could reflect the typical features of the wind farm according to different conditions. Active power, reactive power and the voltage on the grid bus in the wind farm could be used as the typical features of the wind farm, an algorithm based on the above output features is proposed in [9].

3.2. Calculation on equivalent parameters

Several algorithms could be used for the calculation on equivalent parameters. Some methods such as weighted summation have simple calculation process. However, the error of the algorithm is large in large-scale wind farm simulation. In this paper, algorithms on parameter identification based on big data mining such as genetic algorithm, principal component analysis, auto mutation particle swarm optimization algorithm (AMPSO) are discussed; the advantages and disadvantages are shown in table 1.

Table 1. Comparison of the characteristics of each algorithm is in the table below.

Algorithm	Advantages	Disadvantages
Genetic algorithm	String set searching, improve the calculation effect	Less efficient than traditional optimization methods and tend to prematurely converge.
Principal component analysis	Low data dimension, easier identification and interpretation.	Vague meaning of the comprehensive evaluation function when the sign of the factor load of the principal components is positive or negative; Low naming clarity
AMPSO	Faster search, fewer parameters to adjust, simple structure, easy to implement.	Easy to fall into the local optimal solution.

3.2.1. Genetic algorithm. Genetic algorithm is a kind of random searching method derives from the theory of evolution, which includes principles of natural selection and survival of the fittest etc. [9][10].

3.2.2. Principal component analysis. Principal component analysis is a kind of eigenvalues extracting method, which eliminates the correlation between sub vectors, thus removes the axes which contain less information, and reduces the orders of the eigenvalues dimension [11].

3.2.3. AMPSO. Auto mutation particle swarm optimization algorithm take wind turbine patterns as the grouping rule for the wind turbine generator system. The relative parameters are divided into transient and steady-state parameters, and calculated by test measurement and AMPSO algorithm, respectively [12].

IOP Conf. Series: Earth and Environmental Science **133** (2018) 012016 doi:10.1088/1755-1315/133/1/012016

4. Multiple wind turbine dynamic equivalent modelling based on big data mining

If the large wind farm is aggregated to a single wind turbine model, the error would be large. Thus, an equivalent method with a small number of representative wind turbines for dynamic wind farm modeling is proposed and shows a satisfying result in simulation. Some wind turbines have similar operating point can be divided into the same class and represented as an equivalent turbine. This method follows three aspects： cluster analysis of wind turbines, aggregation of parameters in the same cluster and equivalence of collector system. The multiple wind turbine dynamic equivalent modeling flow chart is shown in figure 1.

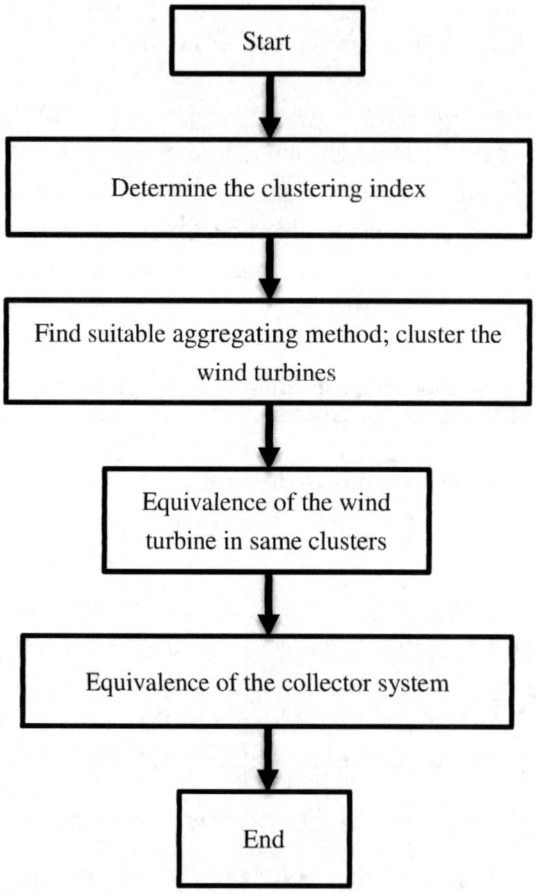

Figure 1. Flow chart of the multiple wind turbine dynamic equivalent modeling.

4.1. Cluster analysis of wind turbines

By suitable index and aggregation algorithm, cluster analysis classifies the wind turbines with the same or similar operating point as one group. Accordingly, a large number of wind turbines can be divided into several groups. The aggregation algorithm is a branch of data mining. It divides physics or abstract sets into many groups, each group also called cluster. Any object in the same cluster has a great similarity while distinction between objects from two clusters is obvious.

IOP Conf. Series: Earth and Environmental Science **133** (2018) 012016 doi:10.1088/1755-1315/133/1/012016

K-means algorithm is one of the traditional aggregation algorithms. Its main idea is to divide objects into clusters through iteration and make the clusters become compact and independent. This algorithm usually ends at local optimal solution.

When Double Fed Induction Generators (DFIGs) have wind speed disturbance or faults at system side, the control system in generators works as before. Ref. [13] picked up the initial value of 13 state variables to reflect starting operating point of generators and used k-means algorithm to aggregate generators. Ref. [14] analyzes the eigenvalues of linear equation set of single machine infinite system with different kinds of generators under different situation, in order to find the leading eigenvalues. Then use the related factors to calculate the related factors between state variables and eigenvalues to find the optimal state variables as cluster index. It reduces the number of cluster index and improves the aggregation speed.

With large-scale wind power generation, a significant performance for the behavior of relay protection and control equipment based on the feature of voltage and current wave has appeared. Ref. [15] established a multiple model for wind farm electromagnetic transient also with k-means algorithm. The cluster index is rotor speed of DFIG at the moment before short-circuit occurs.

Another aggregation method is hierarchical clustering. It builds clusters based on data level to form a tree called clustering tree. Every cluster is a node of the tree. This algorithm can explore data at different granularity level and can achieve the similar measurement and distance measurement easily. However, the biggest problem is its vague stop condition and the method cannot be corrected if combine or split operation has been done. These disadvantages may have an impact on low quality of clustering result. In order to solve the clustering problems, hierarchical clustering is combined with other algorithms.

Ref. [16] applied the hierarchical clustering as the aggregation method, where the characters of the transient voltages were used as the grouping rule of wind turbines. Ref. [17] also identified wind farm groups with similar dynamic behavior by hierarchical clustering. The cluster index is the similarity of low voltage ride through at connecting point in grid-connected wind farms.

According to the advantages and disadvantages of two aggregation methods mentioned above, [18] set up a comprehensive algorithm named H-K clustering. Hierarchical clustering was used to get original information and k-means algorithm to complete the aggregation process. Silhouette value was used as the measurement indicator.

Spectral clustering is an algorithm evolved from graph theory. Compared with the traditional clustering method, spectral clustering has more adaptability to data distribution and better clustering effect, and at the same time, the computational complexity is smaller and the implementation is simpler.

For the wind farms with complex terrain or irregular layout, a spectral clustering method based on the diffusion mapping theory is proposed in [19]. It considered the output properties of active power, reactive power and voltage which combined with the operation environment, wind speed fluctuation, rotate speed and other parameters as the grouping rule.

Using support vector machine as aggregation tool, Support Vector Clustering (SVC) belongs to kernel clustering. It has two significant advantages: first, the cluster boundary can be any shape; second, the noisy data point can be analyzed and the overlapped cluster can be separated while most aggregation methods cannot achieve it.

Ref. [20] proposed a concept of group and divided wind turbines into clusters by SVC according to wind speed. Then divided clusters into groups based on wind direction. Compared with traditional methods, simulation of how wind farm will work in the next whole year can be more accurate.

4.2. Aggregation of parameters in the same cluster

In order to simulate the dynamic wind farm more efficiently in the gird-connected analysis, equivalent generator and its representative parameters which on behalf of generators in the same clusters are needed after the cluster analysis. It is a common method by using volume weighted method and single machine multiplication to get parameters of lumped model [21]. This method has an advantage of easy and less calculation with the actual system model and has a widely practical application. Moreover, the aggregation of parameters in the same cluster can be considered as a single equivalent problem which can be dealt with parameter identification methods.

4.3. Equivalence of wind farm collector system

Conventionally, a large wind farm is made up of dozens or even hundreds of wind turbines which connect to grid through set-up transformers and underground cables collect to the bus. This feature may lead to the more complicated internal wires and may involve a large amount of cable routes. Thus, when it comes to dynamic equivalent analysis, it is necessary to consider the collector system.
Ref. [22] presented an equivalent rule of collector system. Aiming at wind farm detailed parameters, [23] proposed parallel transform method and single impedance equivalent method for the collector system.

5. Conclusion and outlook

To sum up, big data mining method has been widely used in dynamic wind farm modeling up to now, with good results in the fields of wind speed, cluster analysis and aggregation of multiple wind turbine, identification of equivalent parameters for single wind turbine, etc. In the future, operating data will increase as further increase in the scale and number of wind farms. Big data mining method will has a greater potential of researching and applying in dynamic wind farm modeling.

References

[1] GWEC 2016 *J. Global Wind Report.*

[2] Zhang C, Wu X K Review of Dynamic Equivalent Modeling for Wind Farm 2015 *J. Shanxi Electric Power.* **43**

[3] Lin L, Zhao H L, Chen Y, Li D Research Summary of Wind Farm Modeling 2014 *J. Modern Electric Power.* **31**

[4] Li X F, Dong Y F and Li J 2010 *M. Data mining and knowledge discovery.*

[5] Riahinia S, A. Abbaspour, and M.Fotuhi-Firuzabad, et al. A Neural Network-Based Model for Wind Farm Output in Probabilistic Studies of Power Systems 2013 *J. IEEE Electrical Engineering.*

[6] F. González-Longatt, P. Wall and V. Terzija Wake effect in wind farm performance: Steady-state and dynamic behavior 2012 *J. In Renewable Energy.* **39**

[7] Ma Y J, Yang H S, Zhou X S, Li J and Wen H L Modeling of wind farms based on artificial neural networks 2010 *J. Electronic Power.* **43**

[8] Hayes B P, I. Ilie, A. Porpodas, et al. Equivalent Power Curve Model of a Wind Farm Based on Field Measurement Data 2011 *J. IEEE Trondheim PowerTech*

[9] Qiao J G, Lu Z X, Min Y, Liu J, Xie Z J, Wang H Q New Dynamic Equivalence Method for Grid-Connected Wind Farm 2009 *J. Transactions of China Electrotechnical Society*. **24**

[10] Li Hui, Wang H S, Shi X Y and Yang C Study on equivalent model of wind farms based on genetic algorithm 2011 *J. Power System Protection and Control*. **39**

[11] Yu Y, Mi Z Q and Liu X J Modeling Method of Large-scale Wind Farm Based on Operating Data 2011 *J. Acta Energiae Solaris Sinica*. **32**

[12] Li R, Tang F, Liu Y, KE Y and Zhang X Equivalent Model of Doubly-fed Wind Turbine Generator Systems Based on Auto Mutation Particle Swarm Optimization Algorithm 2012 *J. Automation of Electric Power Systems*. **36**

[13] Chen S Y, Wang C, Shen H, Gao N C, Zhu L and Lan H Dynamic Equivalence for Wind Farms Based on Clustering Algorithm 2012 *J. Proceedings of the CSEE*. **32**

[14] Huang W and Zhang X Z Feature Analysis Based Equivalent Modeling for Large-Scale Wind Farms 2013 *J. Power System Technology*. **37**

[15] Zhang B H，Li G H，Wang J，Hao Z G，Zhang J H， Huang R M and Wang X L Impact of wind farm integration on relay protection(2): DFIG-based wind farm electromagnetic transient equivalent model 2013 *J. Electric Power Automation Equipment*. **33**

[16] Xu Y Q and Wang N Study on dynamic equivalence of wind farms with DFIG based on clustering analysis 2013 *J. Journal of North China Electric Power University*. **40**

[17] Fan G Y, Shi K P, Zheng T Y, Feng L M and Li Z Y Cluster Analysis of Grid-Connected Large Scale Wind Farms 2011 *J. Power System Technology*. **35**

[18] Zhu C L， Liu S M， Wang Z J， Pan L and Yin J W Wind Turbine Grouping Method Based on Divisive H-K Clustering Algorithm for Wind Farm 2016 *J. High Voltage Apparatus*.**52**

[19] Lin L and Chen Y Wind turbine grouping with spectral clustering algorithm based on diffusion mapping theory 2013 *J. Electric Power Automation Equipment*. **33**

[20] Ali M, I.Ilie, Jovica V and Milanović Wind Farm Model Aggregation Using Probabilistic Clustering 2013 *J. IEEE Transactions on Power Systems*. **28**

[21] Zhao Y J and Zeng W. Research on Applied Dynamic Equivalence Model for Wind Farm Based on Double-fed Wind Power Generating Units 2012 *J. Guangdong Electric Power*. **25**

[22] Mi Z Q, Su X W, Yang Q X, Wang Y and Wu T Multi-Machine Representation Method for Dynamic Equivalent Model of Wind Farms 2010 *J. Transactions of China Electrotechnical Society*. **25**

[23] Jin Y Q, Huang H, Ju P and Pan X P Collector network transformation methods for wind farm aggregated modeling 2012 *J. Power System Protection and Control*. **40**

The planning and construction of Distributed Energy System in Qingdao Sino-German Eco-park

Cun Wei[1, 2], **Gaijing Zhang**[1, 2] **and Peipei Song**[3, 4]
[1.] Shanghai research institute of building sciences, Shanghai, China.
[2.] 568 Shenfu Rd., Shanghai, China.
[3.] Management Committee of Qingdao Sino-German Eco-park, Qingdao, China.
[4.] 2877 Tuanjie Rd., Qingdao, China.

Abstract. This paper introduce the development and characteristics of new energy, Eco-city and Distributed Energy System in China, a case study of Qingdao Sino-German Eco-park, research on practical application about planning and construction of Distributed Energy System in Eco-city. Results show that: we must first do a good job in energy planning, giving full play to their own advantages, and Distributed Energy System based renewable energy resources is a promising option for reducing emissions from electricity generation in Eco-city.

1. Introduction

At present, the construction of urban has entered a period of rapid development with the increase in the number of urban population. As a result, energy consumption and environment damage have become more and more serious. Now, various initiatives have made great efforts to mitigate them. Along with our country's urbanization process, the energy transformation that takes the new energy as the pivot has become the trend of our country's energy development. With the increasing scale of fossil energy consumption, it is urgent to develop new energy resources and accelerate the transformation of energy. With the new energy as the fulcrum of China's energy transformation system is accelerating change, and vigorously develops new energy has risen to the national strategic level.

For this reason, the Power Generation and Installed Generation Capacity by new energy in China has been rapid development in past the 10 years, which is shown in Fig.1. The Power Generation refers to the capacity of electricity generation in the actual operation; the Installed Generation Capacity refers to the capacity of electricity generation in full load operation.

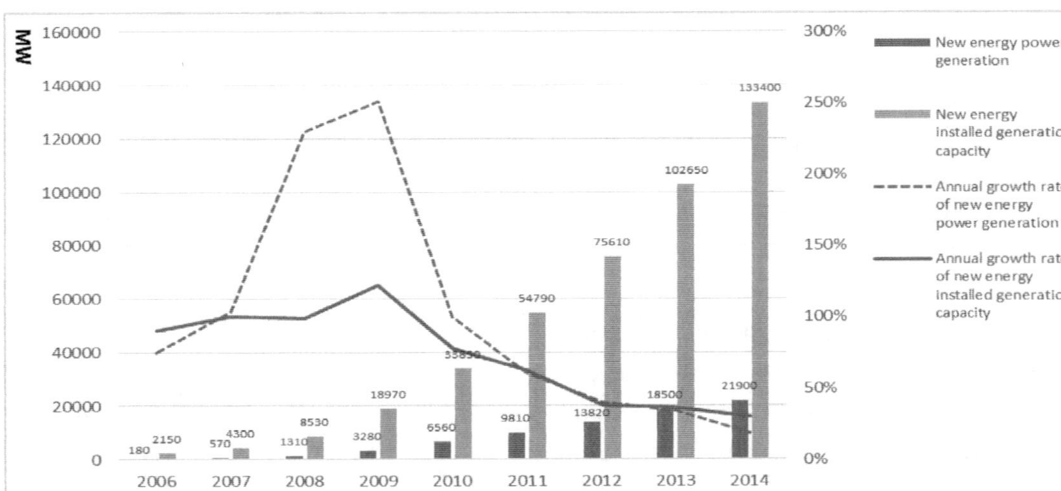

Figure 1. Development of the Power Generation and Installed Generation Capacity by new energy in China

In this situation, energy reform and environment protection have become a new challenge of whole human being. Also, the construction of sustainable developmental harmonious society will ulteriorly accelerate the development and innovation of Chinese urban. The increasing concern for Green, ecological, and low carbon in China is driving government to develop and implement.

Alternative practices especially the construction of Eco-city. The concept of Eco-city is proposed in the process of the "man and biosphere" plan, which was launched by the joint Anglican branch organization in 1971. However, there is no real definition about Eco-city in the world, but the understanding about the intension of Eco-city is also continuing to deepen [1].

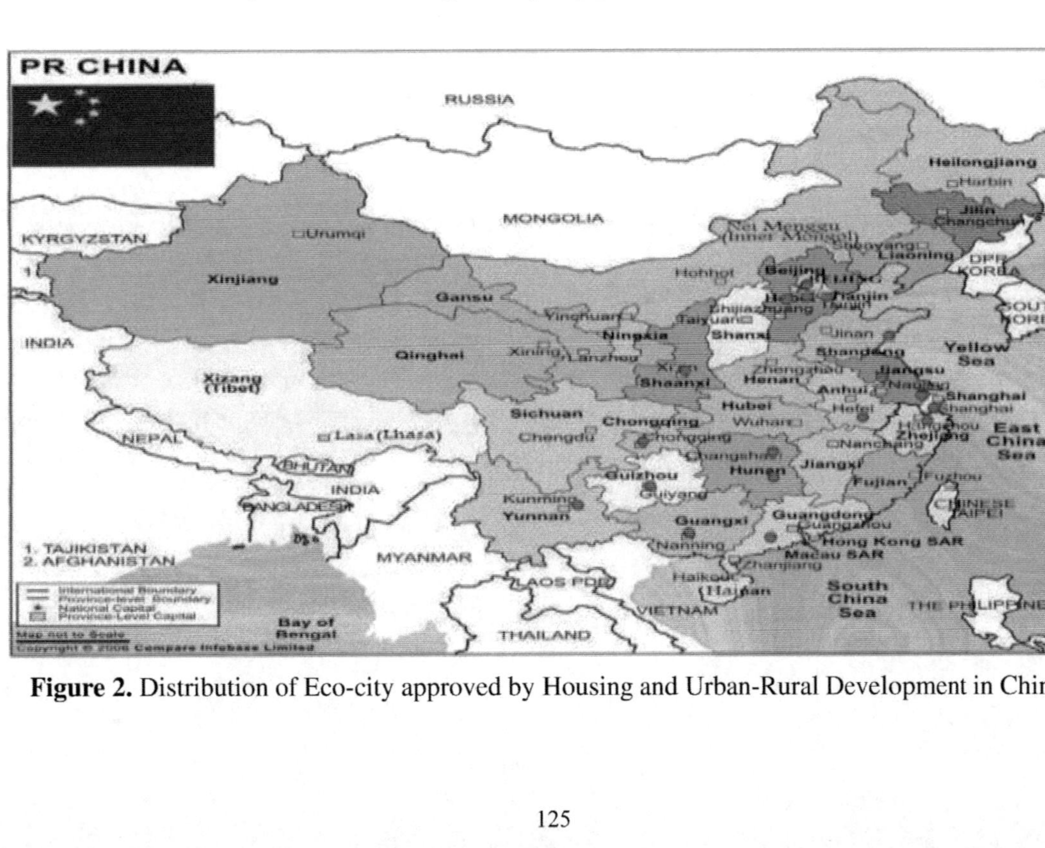

Figure 2. Distribution of Eco-city approved by Housing and Urban-Rural Development in China

IOP Conf. Series: Earth and Environmental Science **133** (2018) 012017 doi:10.1088/1755-1315/133/1/012017

In addition, President of Xi Jinping is proposed to support the construction of Eco-city for improving overall ecological environment quality and low carbon level in the Thirteenth Five-Year Plan of China. Therefore, the number of national Eco-city approved has reached 18 by the end of 2015 including Qingdao Sino-German Eco-park, Tianjin Sino-Singapore Eco-city, etc. [2]. The distribution of national Eco-city is shown in fig.2, in which I simply list these Eco-city approved by Housing and Urban-Rural Development.

2. Development of Distributed Energy in China

By the end of 2014, the Installed Generation Capacity of renewable energy resources has reached 133400Mw, soared by nearly 28% compared with last year, according for 7.9 %of national electricity generation. The installed generation capacity from wind power has reached 95810Mw, according for 27% of global wind power generation, which has exceed other countries all over the world. Similarly, with the rapid growth rate of solar power generation all over the world, the installed generation capacity from solar power reached 26520Mw and ranked second only to the Germany. The Installed Generation Capacity of solar power in most provinces of China at the end of 2014 is shown fig.3.

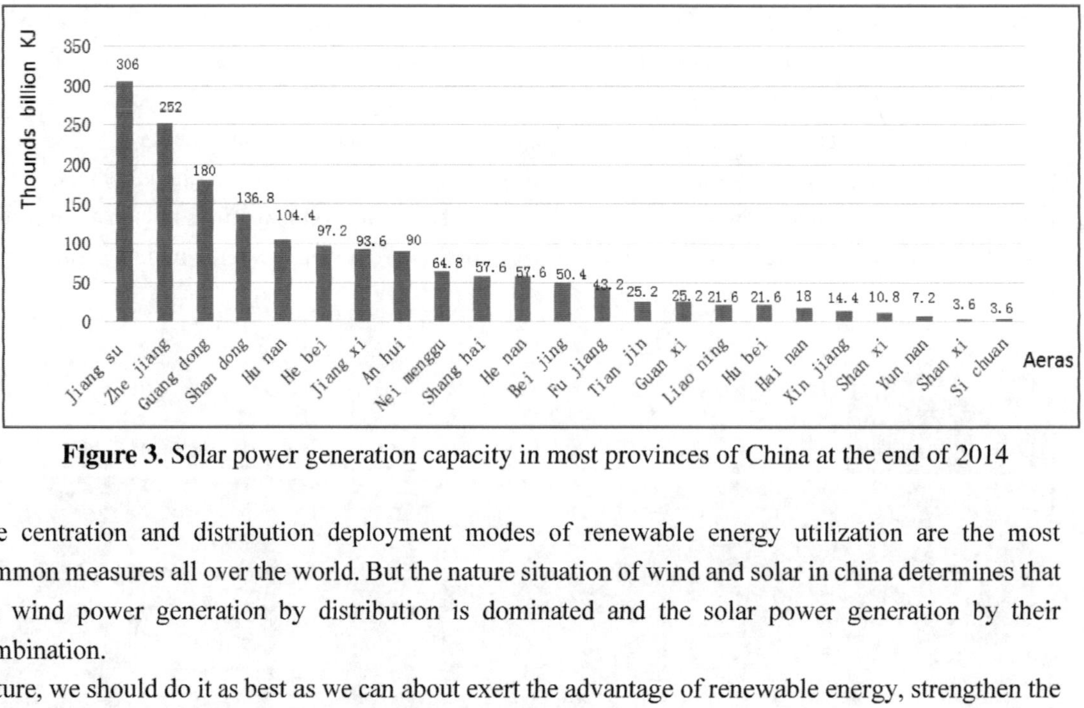

Figure 3. Solar power generation capacity in most provinces of China at the end of 2014

The centration and distribution deployment modes of renewable energy utilization are the most common measures all over the world. But the nature situation of wind and solar in china determines that the wind power generation by distribution is dominated and the solar power generation by their combination.

Future, we should do it as best as we can about exert the advantage of renewable energy, strengthen the development of renewable power generation and the optimization of layout and promote the efficiency of resources transition. Overall objectives of China have been proposed in the "Strategic action plan for energy development (2014-2020)", the Installed Generation Capacity by wind power will reach 0.2 billion Kw, the Installed Generation Capacity by solar power will reach 0.1 billion Kw.

3. The Planning and construction of Distributed Energy System in Qingdao Sino-German Eco-park

Distributed Energy System (DES) refers to local or small-scale energy supply system of independent in order to provide energy for users nearby, including solar power generation, wind power generation, ground-source heat pump and so on in China. DES as a highly efficient energy utilization system is more and more valued by the state and enterprises at home and broad [3-6], which can play an essential role in providing energy with substantial environmental and other benefits especially in Eco-city. This paper introduces a national Eco-city of China, which adopt DES to provide the required energy.

3.1. Introduction of Qingdao Sino-German Eco-park

In order to promote the development of green ecological, German Ministry of economics and technology and the Chinese Ministry of Commerce jointly signed "The Memorandum on the establishment of Qingdao Sino-German Eco-park (QSGE)" during the German Chancellor Angela Merkel's visit to China in July 2010. QSGE is the focus of cooperation projects between China and German, which is also the first joint construction project in the field of Eco-city in China. In addition, President Jia Qinglin, Premier Li Keqiang, Vice Premier Wen Jiabao also mentioned the project of QSGE during the visit to Germany in 2011.The construction of QSGE mainly form parts of the ecological planning, green building, low carbon industry and green management, which involve a total of more than ten aspects, such as index system, green planning system, energy utilization, green municipal, green construction, environmental protection, intelligent development, low carbon industry, urban management, green consulting and so on.

So far, QSGP has received a large number of honorary titles including "National Low Carbon City (Town) pilot", "The First Comprehensive Standard of demonstration park", "National Green Manufacturing International Innovation Park", "The wisdom of the city pilot", "National Eco-city", "New Energy Demonstration Park", Etc. At present, it is still in the stage of construction, the first phase of the planning land area of 11.56 square kilometers. The planning rendering of QSGP is shown in fig.4.

Figure 4. The planning rendering of Qingdao Sino-German Eco-park

3.2. The practice of Distributed Energy System in Qingdao Sino-German Eco-park

IOP Conf. Series: Earth and Environmental Science **133** (2018) 012017 doi:10.1088/1755-1315/133/1/012017

The construction and development of DES in Eco-city depends on the following aspects: first, we should make full consideration about energy consumption and environmental destruction combined with cost estimation and energy technology types in the region while beganing to build the Eco-city, in order to determine the most applicable technology. Secondly, According to the analysis of technologies selected compile the energy planning combined with the size of the region and carry out district construction procedures. In addition, Energy efficiency, the reasonable matching of different energy in the process of using, the circulation and mutual transformation among various energy types from the overall point also should be considered while combine the energy planning.

Based on the above content, and following the design idea of high energy and high use, low energy and low use, peak clipping and filling valley, Complementary guarantee, the DES of QSGE are being set up,which also make the most of renewable and clean energy. What's more, all kinds of energy and resources are allocated reasonably, through the establishment of the recycling system of fossil energy and renewable energy, the combination of centralized and distributed, and the combination of mixing system and independent system, in order to optimize the transmission and distribution system.At the same time, the DES of QSGE also enhances the design of the entire energy supply system fully from the links of production, storage and transportation, application and Regeneration, which not only meet the needs of cold, heat, steam, hot water and some electric load, but also improve the economic indicators of energy utilization, mitigate the impact on the surrounding environment, the DES in QSGP is shown in fig.5.[7]

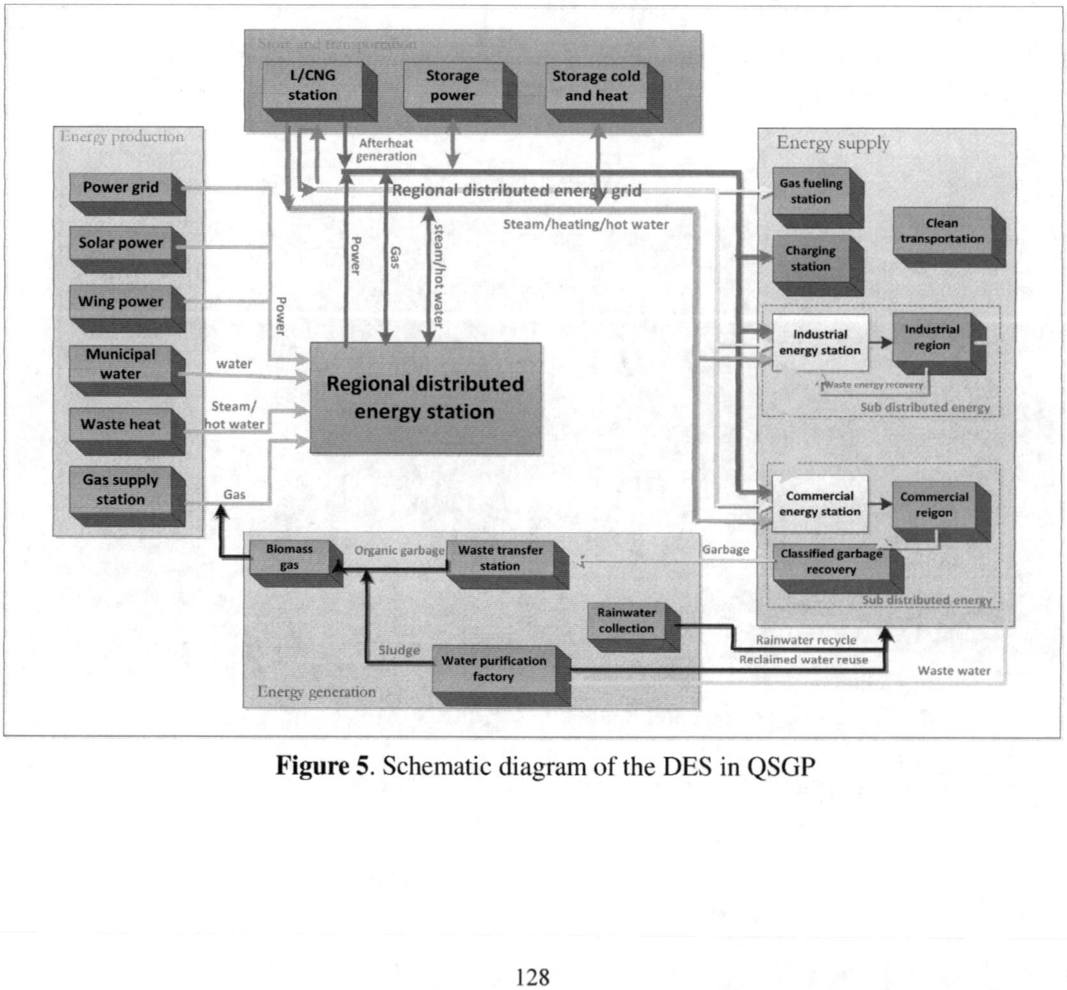

Figure 5. Schematic diagram of the DES in QSGP

IOP Conf. Series: Earth and Environmental Science **133** (2018) 012017 doi:10.1088/1755-1315/133/1/012017

The link of energy production: Using the model of many to single to supply energy, the power, steam, heat water and heat are introduced into the regional distributed energy station. What's more, the power is composed of wind power, solar power and grid. Thus, there is sufficient capacity to provide energy to sub distributed energy and satisfy the requirements of the endpoint.
The link of energy storage and transportation: The station of L/CNG is constructed in QSGP, and the device of storage for cold, heat and power also is constructed in sub distributed energy station. Hence, it not only can play a role of peak saving and valley filling for both electric network and gas network, but also can assure the demand of peak energy while running. As a result, there is a great contribution to reducing the size of the entire energy system and reducing the cost of running in the future.
The link of energy application: At the side of energy supply, it use the method of mutual complement each other in time and space. In other word, we can transport energy form residential areas where demanded seldom energy to commercial and industrial areas where demanded much more energy in the daytime, by contrast in the evening. Meanwhile, for lowering the temperature of the heating medium and improving comfortable quality, Floor Radiant Heating is used in the building. Besides, taking the method of full day monitoring without interruption, the mathematical simulation and analysis of the data were carried out, and then guide the production of energy in the link of energy production. What's more, it must be strictly enforced in the process of design or other national codes during construction.
The link of energy regeneration: It can play a vital role in the recovery of waste heat and energy conservation, by means of recovering the waste heat resource from the regional distributed energy station, sub distributed energy stations and industries. Usually, the recovery of waste heat with low grade needs to increase its temperature or grade before supply to the building or plant by setting the circulating water heat pump system. In addition, it is essential to take effective classification and treatment of resident garbage and wastewater in the link of energy regeneration, which will be used to drive the biomass power generation as raw material and provide organic fertilizer for the nearby farmland.

3.3. The part of renewable in the Distributed Energy System of QSGP

Due to a rich solar energy, wind energy and geothermic in QSGP, renewable and clean energy form the chief institutions of the DES. In solar energy, annual total solar radiation is 5040MJ/m^2, the average annual sunshine hours for 2550 hours. In wind energy,as next to the sea,the annual average wind speed is 5.3m/s, the effective wind energy density is 240.3W/m^3, the effective wind energy average time is 6485 hours every year.In geothermic, sand and stone are the most important part of geology, 0~30 meters is weathered granite, 30~120 meters is sand and stone that comprehensive coefficient of thermal conductivity is 1.464W/m, and the average specific heat capacity is 1.847×10^6J/m^3。Therefore, it has a great advantage in the use of renewable energy in QSGP [8].
According to the Planning of Renewable in QSGP, the renewable energy accounted for the proportion of the total energy consumption of the park plan to reach 15%. In order to achieve the goal, Management Committee of Qingdao Sino-German Eco-park has adopted a series of measures such as solar water heater integrated with building, Building-integrated photovoltaic, Ground-source Heat Pump, Wind power, Biomass-Energy and so on. Detailed instructions as shown in Table 1:

Table 1. The major technologies employed and the predicting of annual supply of renewable

Name	Major technologies employed	Predicted supply	Proportion
Building-integrated solar water heater	More than 90% hot water is provided by the solar hot water system in residential building; More than 50% hot water is provided by the solar hot water system in public building; etc.	4490.0 tce	1.70%
Building-integrated photovoltaic	Use of building-integrated photovoltaic in industrial building; Use the solar energy photovoltaic street lamp; etc.	3433.8 tce	1.30%
Ground-source heat pump	Large-scale promotion of ground-source heat pump system; etc.	1848.0 tce	0.70%
Wind power	The introduction of wind power plant of Zhua Ma Shan; Use the wind power street lamps, etc.	29055.0 tce	11.0%
Biomass-energy	Construction waste treatment gasification station; using bio gas to generate electricity; etc.	1057.0 tce	0.40%
Total	—	39883.8 tce	15%

Based on renewable energy resource, the construction of DES in QSGP has a super advantage compared with the traditional energy systems in terms of energy saving and emission reduction. After preliminary calculation, it can reduce 506 thousand tons of CO_2, 5.8 thousand tons of SO_2, 4.0 thousand tons of NOx, 2.11 thousand tons of dust and so on.

3.4. Investment estimate

The cost of the total investment of the DES in QSGP majorly includes one regional distributed energy station with large scale, nine regional distributed energy station with small scale, six sub distributed energy stations, one operation management center, and transmission and distribution network. Each distributed energy station is made up of factory building, substation room, guards, control equipment, etc.

We determined the cost of major equipment based on the experience, equipment price standard and quotation of equipment manufacturers while estimating the inital investment. The calculation methods of the construction engineering cost are based on the price per subdivision according to its own characterristics and local similar engineering. Based on relevant quotas and budget index, we estimated the cost of installation and combined with local market conditions adjust the price. Detailed costs as shown in Table 2:

Table 2. Estimation of the gross investment about the construction of DES [9]

Code	The name of cost	Investment (Thousand dollars)	Code	The name of cost	Investment (Thousand dollars)
1	Regional distributed energy station	40107.69	13	Intelligent meters	8461.54
2	Sub distributed energy station	53378.46	14	Intelligent control system	6615.38
3	Local distributed energy station	46476.92	15	Energy supervising system	1815.38
4	Auxiliary equipment	6993.85	16	Energy optimization system	2307.69
5	Engineering installation	8287.69	17	Energy regulation system	3846.15
6	CNG storage station	1538.46	18	Energy trading system	1846.15
7	LNG storage station	5230.77	19	Public service system	2307.69
8	Gas fueling station	3384.62	20	Energy cloud computing service system	1153.85
9	Solar water system	12307.69	21	Transmission and distribution network	46153.85
10	Photovoltaic system	23076.92	22	Construction project	7444.62
11	charging station	953.85	23	Land use	8649.23
12	Recycling and utilization system	3076.92		Total	295415.38

4. Conclusion

The planning and construction of Distributed Energy System in QSGP is one of the best successful cases in China. It is worthy of promoting, because the characteristics of high energy utilization, environmental protection and energy conservation.

It is important to take efforts to reduce energy consumption and carbon emission. on the one hand, we must first do a good job in energy planning, giving full play to their own advantages, and Distributed Energy System based renewable energy resources is a promising option for reducing emissions from electricity generation in Eco-city. On the other hand, we should conscientiously implement all planning and policies for promoting energy consumption and carbon emission.

Acknowledgments

This work was financially supported by the National Key Technology R&D Program during the 13th Five-Year Plan Period (2017YFC0702303-03).

References

[1] ShunSheng Wang, ChuanChang Gao. Study about planning and Construction of urban Based on Eco-city Theory. *Applied Mechanics Materials*, Vols71-78(2011) pp220-223.

[2] Xun Li, Bing Li. Development status and trend of green ecological city. China Green Building 2016. *Chinese Society for Urban Studies,* 2016. (In Chinese)

[3] Wang Y L, Shi D, Fan L S, et al. Application and Development of Distributed Energy Systems in China 2015. (In Chinese)

[4] Yang Y W, Yang R F, Ren J X, et al. Analysis of Influences of Residential Energy Prices Change on Development of Residential Distributed Energy System, *Applied Mechanics & Materials*, 2012, 209-211:1852-1857.

[5] Zhou Z, Liu P, Zhang J Y, et al. Evaluating the Impact of Carbon Taxes on the Optimal Design of Distributed Energy Systems. *Advanced Materials Research*, 2012.

[6] Chengzhang Zhu. Measures to Prevent Long-term Oversupplies of Electricity and Natural Gas. *SINO-GLOBAL ENERGY*, 2015. (In Chinese)

[7] Xilai Hu, Distributed Energy Resource Planning in Urban Ecological Characteristic of the Construction Park.Tianjin. The University of Tianjin. (In Chinese)

[8] The Green Planning of Qingdao Sino-German Eco-city. The company of Shanghai research institute of building sciences, 2014. (In Chinese)

[9] The Energy Planning of Qingdao Sino-German Eco-city. The company of ENN group, 2012. (In Chinese)

IOP Conf. Series: Earth and Environmental Science **133** (2018) 012018 doi:10.1088/1755-1315/133/1/012018

Study and design on the typical plan for regional hybrid energy internet

Zhenyu Zou, Shuai Zheng and Zhicheng Sha
Shandong Electric Power Engineering Consulting Institute Corp, Ltd. 106 Minziqian Road, Jinan, Shandong, China

Abstract. The emergence of the energy internet has changed people's understanding of energy production, transmission, storage, conversion, consumption and so on. However, how to promote the development of the energy of the Internet, how to make it with the existing various types of energy entities organic integration, so that it can play an important role, we still need do more in-depth research and practice. After introducing the current situation for energy internet at home and abroad, the definition, character and motion traits of regional energy internet, this paper comprehensively studied typical scheme for regional energy internet based on Multi-energy complementary and developed the feasible suggestion key techniques and directions on development for regional energy internet.

1. Introduction
Energy Internet is a complex multi-net flow system formed by tight coupling with gas networks, transportation networks and other systems based on the Internet and other advanced information technologies, with distributed renewable energy as the main primary energy source and with the power system at the core[1]. The system function, technical framework, information technology and development prospect of the energy Internet are analyzed and expatiated in the documents, respectively, which provides a technical foundation for regional energy management under energy interconnection[2].

Based on the present development of the energy Internet both at home and abroad, this paper presents the definition and characteristics of the regional energy Internet[3], analyzes application patterns of a variety of intelligent control equipment and management technologies by systematically studying the typical plan for the regional energy Internet according to the characteristics of energy flow and information flow based on multi-energy complement, and puts forward a new operating mechanism by summarizing the operation characteristics of the regional energy Internet[4]. In the end, the paper puts forward feasible proposals in respect of the key technologies and directions for future development of the regional energy Internet[5].

2. Regional energy Internet

Energy Internet is a "Wan" of multi-energy information combination built based on the Internet philosophy, correspondingly, the regional energy can be regarded as a "Lan", known as "regional energy Internet". Externally it transfers information with "Wan" and settles energy accounts, and internally it provides energy management and service for regional users[6]. The regional energy Internet is the basis for multi-energy system analysis, and also a concrete embodiment of multi-energy system characteristics[7]. From the perspective of functions, the multi-energy system can integrate various forms of energy organically and regulates distribution of them according to prices, environmental impact and other factors; from the perspective of energy services[8], it can shift electricity from peak periods to off peak periods for reasonably use energy by reasonable regulation with different needs of users taken into account; from the perspective of the energy network, it can promote the development of various energy technologies by collaborative analysis of electrical networks, natural gas networks, heating networks and so on. A region can be as large as a city, town, community, or as small as an industrial park, a large enterprise, building, which in general covers an integrated energy system of power, gas, heating, hydrogen and electrified transportation as well as related communication and information infrastructure, essentially characterized by energy generation, transmission, transformation, storage, consumption and other processes[9]. In this regional network where various energies are integrated, carriers of information include "power flow", "gas flow", "information flow" and "material flow", etc. Due to its small size, the government, energy companies and large industrial enterprises can take the lead to construct and implement the regional energy Internet, and therefore it has higher practical values[10]. As a part of the energy Internet, the regional energy Internet involves a number of energy procedures in different forms and with different characteristics. It includes the energy procedures that can be easily controlled and intermittent ones that are hard to control; includes the energies that are difficult to store in large capacity and the energies that are easy to store and transfer; and there is collaborative supply at the energy generation end and coordinative optimization at the energy consumption end[11].

3. Main characteristic of the regional energy internet

Compared with the main energy Internet across regions, the regional energy Internet takes all local industrial businesses and residents as its user groups and meets the load demand of the local users by data analysis, energy coordination and optimization of the scheduling mechanism based on collection of data on energy production, consumption, transport, storage and other information. Correspondingly, as a link to contact different regional energy Internets, the cross-regional energy Internet realizes long-distance transmission of energy across regions depending on large backbone networks of power transmission, power transmission and other systems, ensuring safe and steady operation of all regional energy Internets within coverage and providing external interfaces when there is energy overflow or gap on the regional energy Internets. In order to fit the energy supply and demand patterns within the region, the regional energy Internet has developed some characteristics differentiating itself from the cross-regional energy Internet based on fully absorbing excellent experience in the Internet development.

3.1. Multi-energy complement

In order to meet the complex load demand of users in the region, the regional energy Internet has distributed a lot of distributed energy facilities within the coverage, with types covering distributed CCHP system, Combined Heat and Power (CHP), photovoltaic power generation, solar collector system, hydrogen generation station, ground source heat pump and other forms, which constitute a

IOP Conf. Series: Earth and Environmental Science **133** (2018) 012018 doi:10.1088/1755-1315/133/1/012018

compound supply system combining power, heat, cold, gas and many other forms of energy, effectively realizing cascaded utilization of energy. Besides, the regional energy Internet provides standard plug and play interfaces for access of all types of distributed energies, but it has also put forward higher request for optimization and control of the energy Internet. To this end, gas-electrical coordination planning, P2G, V2G, fuel cell technologies and other technologies promoting multi-energy integration will play a more important role in the future.

3.2. Two-way Interaction

The regional energy Internet will break the existing source-network-load pattern of energy flow and create a free two-way controllable multi-terminal energy flow pattern, and the distributed energy router will make it possible for energy interconnection of any node within the region. The industry barriers between original heating companies, electricity companies and gas companies will be broken by establishment of energy conversion stations or energy hubs, and residents equipped with distributed generation units are expected to participate in energy supply along with other energy suppliers. In the future, along with the rapid development of the electric car industry, the intelligent electric vehicles centered traffic network will also be integrated into the existing energy Internet mode.

3.3. Full self-government

Different from the traditional energy use patterns, the regional energy Internet makes full use of various energy resources in the region to build a self-sufficient energy system and fully consume distributed energy resources within the region to realize efficient use of various energy facilities. And as an essential part of the main energy Internet, the regional energy Internet maintains a two-way controllable energy flow form with the backbone energy networks and carries on energy and information exchanges with other regional energy Internets through the large backbone energy networks.

In conclusion, the regional energy Internet is mainly characterized by using the "Internet plus" thinking to reset the energy network needs and achieve a high degree of integration of energy and information, so as to promote the construction of informatization infrastructure of energy networks. By introducing online trading platforms, big data processing and other technologies, the energy Internet will fully tap data on energy production, transport, consumption, transformation, storage and other information, and explore technologies by means of energy demand forecasts, demand-side responses and other information to guide energy production and scheduling.

4. Typical plan for the regional energy Internet

4.1. Overview of the planned region

The region is planned to cover an area of about 64 square kilometers in total, which will be constructed along with new-type urbanization in combination with comprehensive management of canal ecology, building all unincorporated villages into creative towns. So far 1100 mu of the land has been constructed and used, with buildings covering an area of 540,000 square meters. The new land planned for construction will be 2200 mu and the new construction area will be 1.5 million square meters by 2020, mainly for shopping center, maker base, film and TV production base, elderly care and health care base, spa resort, education and other industries. According to the comprehensive development plan for the near future and energy use characteristics in different functional areas, the regional maximum power load is measured 30MW.

4.2. Study on the plan

To build the intelligent energy system for the planned region according to the collaborative development idea of "source-network-load and Internet plus", based on planning and load demand characteristics of the entire 64 square kilometre region, the whole regional energy system includes energy stations (distributed rooftop PV – PV town, farmers light complementary, biomass, waste and biogas power generation for centralized heating, distributed gas CCHP, LNG vaporizing stations), energy networks (integrated tube galleries in the park, connected through power, heating and gas networks between energy stations and linked together with the user load), integrated energy control centres--building an integrated energy control system based on information networks, big data, cloud computing, the Internet and other technologies, to realize order form-based production, intelligent visualized selective consumption, multi-dimensional optimal allocation of energy production elements; employing distributed centralized heating to make gas, electricity, heating, cooling, water and other integrated energies couple with each other, thus improving the efficiency of energy utilization.

The original unincorporated villages will be merged into 14 new featured towns. And all new buildings in the towns will be combined with PV organically, maximizing the use of renewable energy. The planned PV installed capacity is up to 25MW. And each town will be equipped with a gas CCHP energy station, realizing centralized heating and cooling supply in the neighbourhood to reduce energy loss. The surrounding areas of villages and town will be arranged with PV units to improve the efficiency of land use; use of electric vehicles will be promoted and integrated charging piles will be installed in the park to reduce greenhouse gas emissions while consuming the renewable energy.

The plan for the near future is developed around the existing region and the area to be developed before, a total of 3300 mu, planning to supply heating and cooling and provide hot water for life based on the advanced air-source heat pumps by using the distributed energy for natural gas engine CCHP distributed energy at the core, supplemented by distributed PV and wind energy systems, linked by power, shift electricity from peak periods to off peak periods based on energy storage units (including cool storage, heat storage, electricity storage), and optimize the use of energy by storing cool and heat with valley point electricity; the supporting works include distribution networks, comprehensive pipe racks, sewage treatment, building of car charging piles, integrated street lamps with scenery storage and integrated energy management centre; and complete intelligent energy use and measurement units along with completion of the buildings.

4.2.1. Multi-energy complement. According to energy requirements in 2020, with construction of the gas CCHP system at its core, the plan seeks comprehensive utilization of various renewable energy sources (solar power, wind energy, air source heat pumps, integrated street lamps with scenery storage), takes advantage of off-peak power at night to reduce the difference between peak and valley electricity, regulates changes in load of users by means of cool storage, heat storage and power storage, thus constituting an integrated intelligent energy system of multi-energy complement of primary and secondary energies within the region, to realize green, efficient and economical energy use.

According to the preliminary estimate, annual consumption of natural gas will be 40 million Nm3, 5.5 million kWh of solar power, 800,000 kWh of wind power, and 600,000 kWh generated by landscape lamps. Due to the massive use of air source heat pumps and energy storage for energy supply, the proportion of non-fossil fuel energy supplies will be over 30%.

4.2.2. Integration and optimization. By comprehensive intelligent use and management of multiple energy resources by means of "Internet plus", energy conversion stations can be placed in the in the neighborhood according to different load demands of distributed users for integration and optimization by interconnection between power, heating and gas networks. A reasonable and economical operation plan should be developed by installing meters at the user end and analyzing the energy use habits of the users based on big data. In terms of energy production, solar and wind power as well as other renewable sources of energy will be used as preferences for heating supply within the region, with the shortfall to be supplemented by gas or power, besides, cool storage, heat storage, power storage street lamps, and other energy storage ways will be employed to maximize development and use of the renewable energy and increase the consumption proportion of the renewable energy.

A total of 28MW internal combustion engine generator set will be built at the energy station, fume waste heat and jacket heat after natural gas combustion and utilizing will enter the lithium bromide unit for further use, supplying cool in summer and heat or domestic hot water in winter after heat transfer, to realize cascaded utilization of energy. The overall efficiency for energy utilization at the energy station is about 84.3%.

As the day is a peak time in cities, PV power can complement urban consumption, helping reduce gas consumption of the energy station. The project gives priority to the use of renewable energy, which will be fully consumed if energy storage units are used in combination.

By centralizing the management of energy supply and building an integrated energy management system, the basic parameters of users can be automatically transmitted to the energy monitoring center through the remote monitoring device. The center will collect and analyze the data and regulate the operating parameters of the system in time. With accumulation of historical data, the system can forecast energy use, facilitating energy scheduling preparation. An integrated supply-purchase-sale electricity information acquisition system measuring in real time on demand will be built for construction of the power service platform, to build an intelligent and green interactive service system.

4.2.3. System reliability. Because the system is based on grid-connected operation, it can guarantee the sustainable and stable supply of electricity and other energy sources. If the external power supply is interrupted at night due to any special circumstance, the internal combustion engine generator set can be ensured to start quickly (less than 30 minutes from start to full load), providing power protection for the park. The energy station is equipped with two gas sources. There is pipeline gas connecting the station besides its own LNG storage tanks and gasification plant. Even if the two gas sources are at failure, the power from the empirical grid will ensure cold and heat supply in the park. The energy supply reliability rate is over 99.5%.

4.2.4. System intelligence. The integrated energy management system is a large open platform for energy management services supported by big data, Internet of things, mobile Internet and other technologies. The system is in a hierarchical distributed structure and performs intelligent processing of all information it "has sensed" on production, transmission and consumption of power, heat (cool), water, gas and other energy with the help of the cloud data center , to monitor and manage the whole energy system.

The integrated energy management system can change energy users' passive consumption in the planned mode, and can provide a platform for implementation of the market-oriented energy pricing mechanism, so that energy market players can be energy producers and consumers and actively engage themselves in balancing energy supply and demand, realizing interactions between different energy producers, interactions between different consumers and interactions between producers and consumers

5. Summary

The typical plan presented in this paper creates a solution that is based on cooperation and joint development between energy suppliers and users, subject to sharing revenues and risks. Technically, natural gas, solar and wind power as well as other clean energy are used to replace coal, with cascaded utilization of the gas turbine waste heat, utilization of waste energy of air source heat pumps, integrated innovation of cool (heat) storage units, having different types of energy complement each other, to improve the overall efficiency of energy utilization, create clean energy demonstration zones, and employ big data, cloud computing and other advanced Internet technologies in combination with intelligent energy control strategies to control energy use. It will provide a strong demonstration effect to optimizing urban energy structures, full exploitation of renewable resources, increasing the proportion of energy consumption of gas and electricity, achieving secured and reliable energy supply, diversifying and cleaning energy supply patterns.

Energy construction and development of the energy Internet will be a long and gradual process, and the plan put forward in this paper will be also constantly modified and improved in practice. Construction of the energy Internet needs to be advanced jointly by physical energy systems, information and communication systems, optimal scheduling systems and auxiliary control systems, which complement each other and need harmonious development.

References

[1] Zhaoyang Dong, Junhua Zhao, Fushuan Wen, et al. From smart grid to energy internet: basic concept and research framework , *J.Automation of Electric Power Systems*, vol. 38, no. 15, pp. 1-10, Aug. 2014.

[2] Hongbo Ren, Jian Yang, Yinyin Ban, et al. Exploration and Prospect of distributed energy Internet, *J. Energy of China*, vol. 37, no. 3, pp. 38-42, Mar. 2014.

[3] Yang Li, Ming Wu, Haiming Zhou, et al. Study on Some Key Problems Related to Regional Multi Energy System Based on Universal Flow Model，*J. Power System Technology*, vol. 39, no. 8, pp. 53-58, Aug. 2015

[4] Haiming Zhou, Guangyi Liu, Chaoqun Liu, et al. Study on the Energy Internet Technology Framework, *J.Electric Power* , vol. 47, no.11, pp. 140-144, Nov. 2014

[5] Jun Lu, Shaoliang Zhang, Liu Yang. Discussion on Optimization Design and Planning of Regional Energy Project, *J. Chongqing Architecture*, vol. 13, no. 10, pp. 5-8, Oct. 2014

[6] Hongjie Jia, Dan Wang, Xiandong Xu, et al. Research on Some Key Problems Related to Integrated Energy Systems, *J.Automation of Electric Power Systems*, vol. 39, no.7, pp. 198-207, Feb. 2015

[7] Dong Wang, Yongbiao Yang, Shengjun Yan, et al. Collaboration Mechanism of Management and Control Instructions Among Multiple Energy Systems of Intelligent Park, *J. Automation of Electric Power Systems*, vol. 40, no. 24, pp. 21-26, Dec. 2016

[8] Qadrdan M, Abeysekera M, Chaudry M, et al. Role of power-to-gas in an integrated gas and electricity system in Great Britain. *J.International Journal of Hydrogen Energy*, vol. 40, no. 17, pp. 5763-5775, Sep. 2015

[9] Jiye Wang, Kun Meng, Junwei Cao, et al. Information Technology for Energy Internet, *J. Journal of Computer Research and Development*, vol. 52, no. 5, pp. 1109-1126, May. 2015.

[10] Fang Yang, Cuifen Bai, Yibin Zhang. Research on the Value and Implementation Framework of Energy Internet, *J. Proceedings of the CSEE*, vol. 35, no. 14, pp. 3495-3502, Jul. 2015.

[11] Dongxia Zhang, Xin Miao, Liping Liu, et al. Research on Development Strategy for Smart Grid Big Data, *J. Proceedings of the CSEE*, vol. 35, no. 1, pp. 2-12, Jan. 2015.

IOP Conf. Series: Earth and Environmental Science **133** (2018) 012019 doi:10.1088/1755-1315/133/1/012019

Assessment of wave energy potential along the south coast of Java Island

Qingyang Song and Roberto Mayerle
FTZ-Westküste/Coastal Research Laboratory, Christian-Albrechts-Universität zu Kiel, Kiel, Germany.

Abstract. The south coast of Java Island has a great potential for wave energy. A long-term analysis of a 10-year wave dataset obtained from the ERA-Interim database is performed for preliminary wave energy assessment in this area, and it was seen that the annual median power is expected to exceed 20kW/m along the coast. A coastal wave model with an unstructured grid was run to reveal the wave conditions and to assess the wave energy potential along the coast in detail. The effect of swells and local wind on the wave conditions is investigated. Annual median wave power, water depth and distance from the coast are selected as criteria for the identification of suitable locations for wave energy conversion. Two zones within the study area emerge to be suitable for wave energy extraction. Swells from the southwest turned out to be the major source of wave energy and highest monthly median wave power reached about 33kW/m.

1. Introduction

Wave energy has recently become the interest of many researchers around the world. Several studies on global wave resource based on the data obtained from global wave models such as ECMWF (European Centre for Medium-Range Weather Forecasts) and NWW3 (National Oceanic and Atmospheric Administration WAVEWATCH III) reveal that higher wave heights are more likely to appear on the east coast of oceans in both hemispheres [1]. Based on the wave energy atlas [1], the highest power for primary swell appears off the western Australian coast, reaching an average of 60kW/m during the winter months of the southern hemisphere. Since the main direction of these swell are northeast, they reach the Java Island with considerable wave energy. Another advantage of exploiting the wave energy in this area is that the seasonal variation of wave conditions is much lower than in most areas of the Northern Hemisphere [2]. Thus, wave converters can give more stable output power throughout the year and the lower appearance of extreme conditions can reduce the costs of wave projects.

Site selection for the installation of wave energy devices depends on available power, site characteristics, and possible environmental and socio-economic impacts of the projects. However, wave conditions are still decisive criteria [3]. Several studies have successfully applied wave models to assess the wave energy potential at different locations in Europe and North America utilizing the wave data covering around 10 years from WAM or WAVEWATCH III to analyze the long-term wave variability and to roughly estimate exploitable wave energy [4], [5], [6], [7]. Coastal wave models with refined computational grids and more accurate bathymetric data were also put in progress to determine optimum locations for a wave farm. In this study, results of an assessment of the wave conditions and the wave energy potential in the south coast of Java Island are presented.

2. Methodology and application

Within the investigated domain in this study, waves produced in the Indian Ocean spreading to the south coast of Java Island, and local winds, as a part of the monsoon system, are main sources of wave energy. Therefore, different wave systems due to swell and local winds can be expected, and separate parameters for the wave systems allow estimation of the power from the dominant wave components using respective peak periods. Thus, precise information can be provided for planning and design of wave farms [1]. The year-to-year variability of the wave conditions is revealed by long-term analysis. After that, the detailed seasonal variations and the wave energy in coastal area are presented through wave simulations.

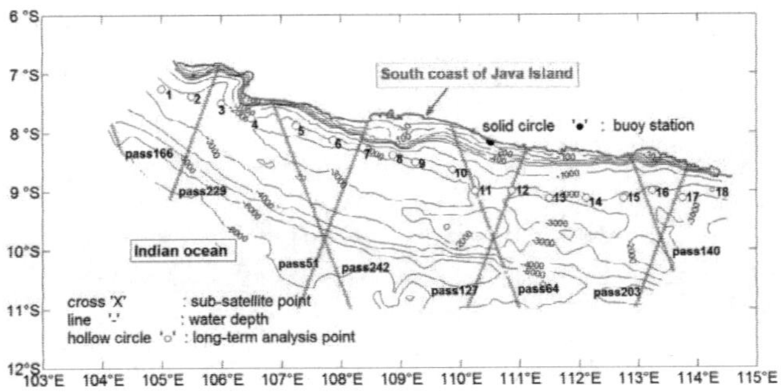

Figure 1. Study domain with bathymetry and locations of satellite tracks, wave buoy station and long-term analysis points.

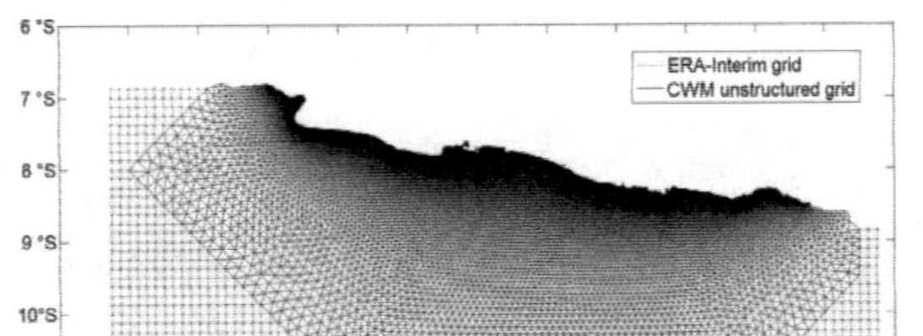

Figure 2. Unstructured grid of CWM in the computational domain.

The ERA-Interim dataset from ECMWF is chosen for the long-term wave analysis. It provides 6-hourly values on significant wave height, mean wave period and wave direction [8]. The wave conditions in deep water along the coast using wave data covering 10 years (2004-2014) are analyzed to roughly evaluate the wave potential in down-wave directions. For the analysis, 18 points with

IOP Conf. Series: Earth and Environmental Science **133** (2018) 012019 doi:10.1088/1755-1315/133/1/012019

constant distances along the 2000m isobath line off the coast are selected (Figure 1). Besides, the wave data from ERA-Interim are also utilized to drive the coastal wave model.

Numerical wave simulations are performed with SWAN (Simulating Waves Nearshore) [9] to reveal detailed wave conditions near the coast. An unstructured triangle grid is constructed and applied to keep the high resolution in the coastal area, and avoiding unnecessary computation resource cost in deep water (Figure 2). The coarsest meshes are on the open boundary matching the grid of the ERA-Interim dataset with a resolution around 14km, while the finest meshes are off the coast with an average resolution of about 500m. The criteria based on a 2D spectrum, which have been used also in WAM, are implemented in SWAN for the wind-swell-wave identification [10].

Coastal wave model (CWM) is validated based on both wave buoy and satellite data. Location of the wave buoy, which generates raw north, west and vertical displacements at a rate of 1.28 Hz, is indicated in Figure 1. Figure 3a and Figure 3b show the comparisons of significant wave height and mean wave period between CWM and buoy observations, and the comparison of significant wave height due to swell (Hsswell) and local wind (Hswind) can be seen in Figure 3c and Figure 3d. Figure 3c and Figure 3d also reveal that the main wave energy comes from swells at the location of the buoy.

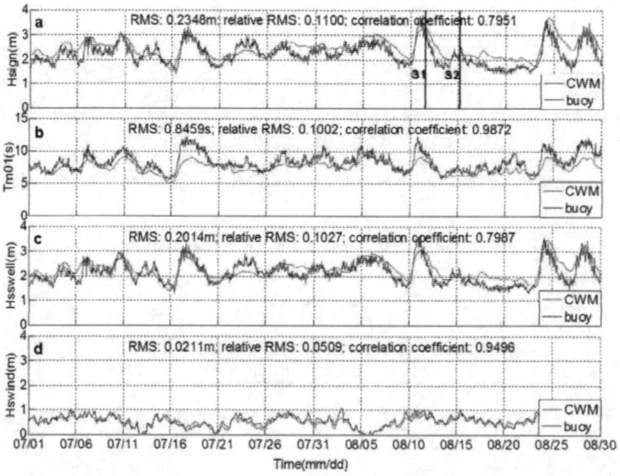

Figure 3. Comparison of time series between CWM and buoy: (a) significant wave height; (b) mean wave period; (c) significant wave height due to swells; (d) significant wave height due to local winds.

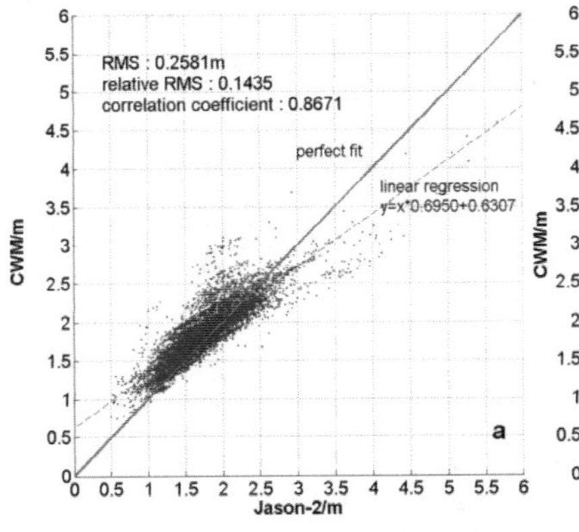

Figure 4. Significant wave height scatter plot of CWM (coastal wave model) versus altimetry data from Jason-2: (a) rainy season; (b) dry season

The altimeter data from Jason-2, provided by AVISO (Achieving Validation and Interpolation of Satellite Oceanographic Data), is also employed to validate the model due to lack of in-situ wave measurements along the south coast of Java Island. The dataset contains significant wave height only. The measurement interval at one sub-satellite point is about 10 days. The locations of sub-satellite points on 8 passes in the model domain are showed in Figure 1. Figure 4a and Figure 4b show significant wave height scatter plot of CWM versus altimetry data from Jason-2. The figures show that CWM has a better agreement with altimetry data in the dry season (2014/5 to 2014/10) than in the rainy season (2013/11 to 2014/04). It is seen that the difference of significant wave height between CWM and the altimetry data are usually smaller than 0.5m.

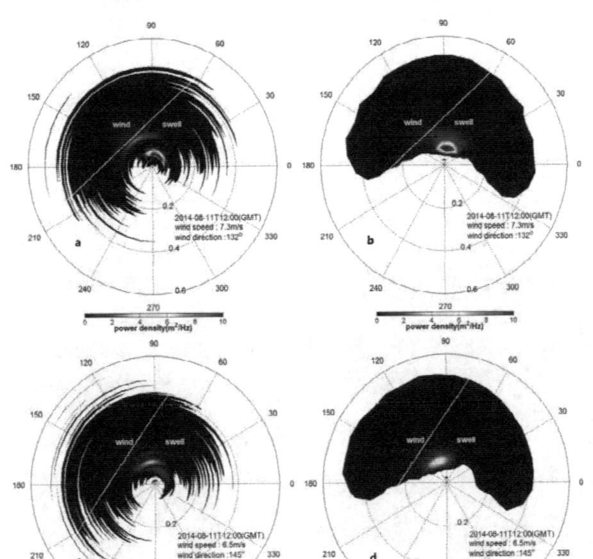

Figure 5. 2D spectrum comparison between CWM and buoy: (a, c) buoy; (b, d) CWM.

IOP Conf. Series: Earth and Environmental Science **133** (2018) 012019 doi:10.1088/1755-1315/133/1/012019

Figure 5a-d reveals that swells dominate the wave field both in extreme wave condition (Figure 5a, Figure 5b) and the average wave condition (Figure 5c, Figure 5d). Primary swells in two time points (S1, S2 in Figure 3a) are from different directions. In the southern hemisphere, the strong westerlies result in heightened wave activities in mid-latitude throughout the year and those waves combine with the trade wind waves when propagate to north [1]. This fact implies that the southwesterly primary swell in the extreme wave condition is due to the westerlies while the southeasterly primary swell in the average wave condition is due to the trade wind.

3. Characterization of wave energy potential in the study area

Based on the 10-year wave data from the 18 points along the 2000m isobath line off the coast, Mean significant wave height is about 2.04m and the mean wave period is about 10.69s. The inter-annual variation of these two wave parameters is small. In addition, significant wave height in dry seasons is higher than in rainy seasons with a mean difference of about 0.43m. Furthermore, mean wave period in dry seasons is approximately 0.4s higher than rainy seasons. Since the points for the analysis are located in deep waters, mean annual median wave power can be estimated with the wave statistic parameters as follows:

$$P = \frac{\rho g^2}{64\pi} H_S^2 T_{m01}\, H_s \tag{1}$$

where ρ is water density, g is gravitational acceleration, H_s is significant wave height, T_{m01} is mean wave period. Estimated mean annual median wave power based on Eq. (1) is 22kW/m, while required annual median wave power for most commercial wave energy converters is around 20kW/m [2]. The seasonal variation of wave power in this area is insignificant. In rainy seasons, the seasonal median wave power reaches 17kW/m and in dry seasons, the seasonal median wave power is 27kW/m (Table. 1).

Table 1. Temporal variations of average wave statistic parameters from 2004 to 2014.

year	Hs (m)			Tm01 (s)			Power (kW/m)		
	An.	dry	rainy	An.	dry	rainy	An.	dry	rainy
04-05	2.08	2.29	1.87	11.0	11.2	10.9	23.5	28.9	18.7
05-06	2.13	2.33	1.93	10.9	11.1	10.8	24.5	29.7	19.8
06-07	2.08	2.33	1.83	10.8	11.1	10.6	23.0	29.5	17.3
07-08	2.10	2.28	1.91	10.6	10.8	10.4	22.9	27.7	18.7
08-09	2.05	2.25	1.85	10.7	11.0	10.4	22.0	27.3	17.5
09-10	1.89	2.14	1.64	10.5	10.8	10.3	18.5	24.3	13.7
10-11	1.98	2.13	1.84	10.3	10.6	10.0	19.9	23.6	16.6
11-12	2.02	2.19	1.85	10.4	10.5	10.3	20.9	24.9	17.4
12-13	2.07	2.28	1.85	10.8	10.9	10.6	22.7	27.9	17.9
13-14	2.04	2.34	1.74	10.8	10.9	10.7	22.2	29.4	16.0
mean	2.04	2.26	1.83	10.7	10.9	10.5	22.0	27.3	17.4

Detailed information of wave conditions in the coastal area are based on the results from CWM. Wave power calculations are made using equation:

$$P = \int_0^{2\pi} \int_{f_{min}}^{f_{max}} \rho g c_g E(\theta, f) df d\theta \tag{2}$$

where P is the wave power, c_g is the wave group speed and $E(\theta, f)$ is the power density as a function of the wave direction (θ) and frequency (f).
Monthly median significant wave height, mean wave period and wave power within 30 nautical miles off the coast are estimated. Swells provide the majority of wave energy (Figure 6a), while mild local winds of speed about 3-5m/s does not play an important role on wave conditions through the year. The average monthly median wave power for swells is estimated as 15kW/m from November 2013 to April 2014, and 25kW/m from May 2014 to October 2014. Meanwhile, the wave power for winds is less than 1kW/m through the year and varies with the wind speed and direction. With the dominance of the swells, significant wave height in the dry season is higher than in the rainy season. The mean wave period in this area is around 6s-8s through the year. The stronger swells in dry season result in the larger mean wave period of around 8s. The peak wave power appears in August 2014 during the period. The detailed information in that month is provided by Figure 7. In August 2014, the monthly median wave power of more than 30kw/m is mainly from the southerly swells. The local winds from southeast only provide the wave power of 0.9kw/m.

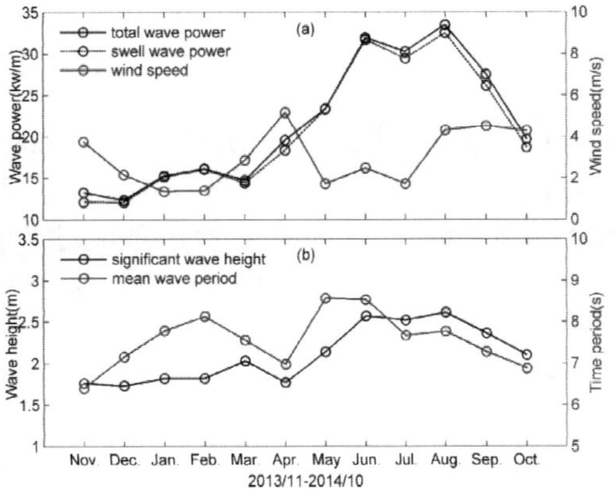

Figure 6 (a). Seasonal variations of monthly median wave power, wave power due to swells and wind from 2013/11 to 2014/10 over the area within 30 nautical miles off the coast. (b) Seasonal variations of monthly median significant wave height and mean wave period.

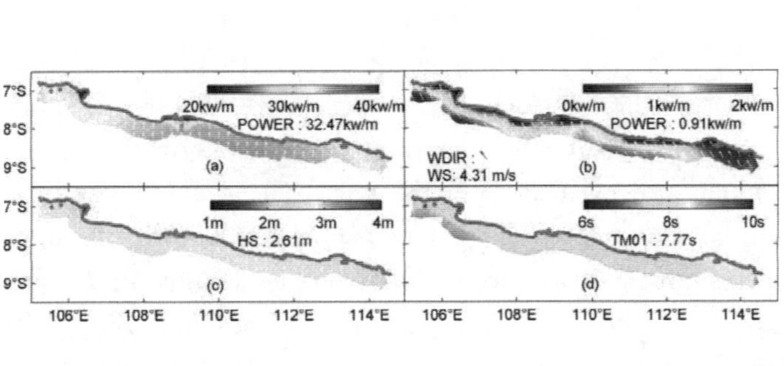

Figure 7. Monthly median wave conditions in Aug. 2014 within 30 nautical miles off the coast. (a) Wave power and energy transport direction due to swells; (b) Wave power and energy transport

direction due to local winds; (c) Significant wave height and mean wave direction2014; (d) Mean wave period. POWER: average wave power over the area; WDIR: average wind direction over the area; WS: average wind speed over the area; HS: average significant wave height over the area; TM01: average mean wave direction over the are

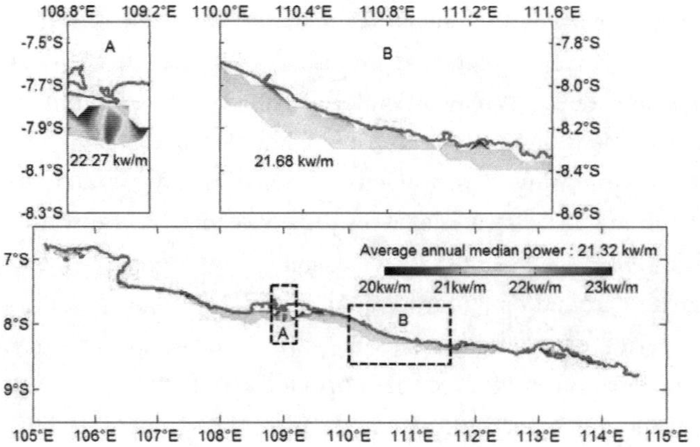

Figure 8. Distribution of Annual median wave power along the south coast of Java Island

4. Site Selection

The selection of locations for deployment of wave energy farms or converters is an issue involving wave conditions, technical feasibility, economic benefit and environmental impact. Three simple criteria for preliminary screening of locations suitable for wave energy farms are presented as follows: 1. Annual median power is larger than 20kW/m; 2. Water depth is less than 200m and greater than 30m; 3. Distance to the coast is less than 12 nautical miles.

Based on the described criteria and one-year output from CWM, regions unfit for wave energy farms are filtered. Figure 8 shows suitable locations, which appears mainly off the middle part of the coast, concentrating on two zones, marked by dash lines.

Since the swells is more likely to approach the eastern and middle part of the Java Island at first, annual median power in most areas off the western part of the coast is lower than 20kW/m, insufficient for wave energy farms. On the other hand, water depth off the eastern part of the coast is deep, more than 200m in most areas. Mooring wave energy converters in that location is more difficult and more costly than in shallower water, although the annual median power is acceptable.

5. Summary

Long-term analysis of wave data from ERA-Interim over 10 years at the 18 points along the isobath of 2000m suggests that there is considerable wave energy available off the south coast of Java Island. Detailed wave conditions are revealed from the buoy data and the coastal wave model, CWM. Generally, the majority of wave energy in this area is provided by swells from southwest produced in mid-latitude region. Thus, the swells as well as the wave power due to the swells in the study area strengthen or recede along with the seasonal variation of the westerlies. From 2013/11 to 2014/10, the highest monthly wave power of 31kW/m appears in Aug., 2014. Additionally, from the 2D spectrums of the buoy data and the wave case studies, another wave system due to southeasterly swells is found and dominates the wave field in average wave conditions in the dry season. Although the mild southwesterly trade wind of speed about 3-5m/s through the year only have limited local impact on waves, the waves will grow through the fetch and perform as relative weak southeasterly swells when approach the study coastal area. After setting annual median wave power, water depth, and distance from the coast as criteria, two zones suitable for wave energy farms emerged. Zone A has the highest annual median power around 23kW/m. Zone B has more suitable locations where some of them are close to the coast. But for further screen the suitable locations for deployment of wave energy farms, more detailed measurement and associated environmental and economical evaluation are required. In the future, assimilation method can be applied to further improve the model results and help to perform detailed evaluations for specific wave energy converters.

References

[1] Arinaga, R. a., & Cheung, K. F. (2012). Atlas of global wave energy from 10 years of reanalysis and hindcast data. *Renewable Energy*, 39(1), 49–64.

[2] Falcão, A. F. D. O. (2010). Wave energy utilization: A review of the technologies. *Renewable and Sustainable Energy Reviews*, 14(3), 899–918.

[3] Nobre, A., Pacheco, M., Jorge, R., Lopes, M. F. P., & Gato, L. M. C. (2009). Geo-spatial multi-criteria analysis for wave energy conversion system deployment. *Renewable Energy*, 34(1), 97–111.

[4] Henfridsson, U., Neimane, V., Strand, K., Kapper, R., Bernhoff, H., Danielsson, O., … Bergman, K. (2007). Wave energy potential in the Baltic Sea and the Danish part of the North Sea, with reflections on the Skagerrak. *Renewable Energy*, 32(12), 2069–2084.

[5] Iglesias, G., & Carballo, R. (2010). Wave energy resource in the Estaca de Bares area (Spain). *Renewable Energy*, 35(7), 1574–1584.

[6] Iglesias, G., López, M., Carballo, R., Castro, a., Fraguela, J. a., & Frigaard, P. (2009). Wave energy potential in Galicia (NW Spain). *Renewable Energy*, 34(11), 2323–2333.

[7] Stopa, J. E., Cheung, K. F., & Chen, Y. L. (2011). Assessment of wave energy resources in Hawaii. *Renewable Energy*, 36(2), 554–567.

[8] Dee, D. P., Uppala, S. M., Simmons, a. J., Berrisford, P., Poli, P., Kobayashi, S. Vitart, F. (2011). The ERA-Interim reanalysis: Configuration and performance of the data assimilation system. *Quarterly Journal of the Royal Meteorological Society*, 137(656), 553–597.

[9] Booij, N., Ris, R. C., & Holthuijsen, L. H. (1999). A third-generation wave model for coastal regions: 1. Model description and validation. Journal of Geophysical Research, 104(C4), 7649.

[10] Komen, G. J., L. Cavaleri, M. A. Donelan, K. Hasselmann, S. Hasselmann and P. A. E. M. Janssen (1994). *Dynamics and Modelling of Ocean Waves. Cambridge University Press*, 554 pp..

IOP Conf. Series: Earth and Environmental Science **133** (2018) 012020 doi:10.1088/1755-1315/133/1/012020

Application of Islanding Detection and Classification of Power Quality Disturbance in Hybrid Energy System

L B Sun, Z S Wu and K K Yang

School of electrical engineering, Beijing Jiaotong University, Beijing 100044, Peoples R China

Abstract. Islanding and power quality (PQ) disturbances in hybrid energy system become more serious with the application of renewable energy sources. In this paper, a novel method based on wavelet transform (WT) and modified feed forward neural network (FNN) is proposed to detect islanding and classify PQ problems. First, the performance indices, i.e., the energy content and SD of the transformed signal are extracted from the negative sequence component of the voltage signal at PCC using WT. Afterward, WT indices are fed to train FNNs midfield by Particle Swarm Optimization (PSO) which is a novel heuristic optimization method. Then, the results of simulation based on WT-PSOFNN are discussed in MATLAB/SIMULINK. Simulations on the hybrid power system show that the accuracy can be significantly improved by the proposed method in detecting and classifying of different disturbances connected to multiple distributed generations.

1. Introduction

Under the new situations and circumstances of a worldwide avocation of energy conservation and reduction of pollutant emission, the renewable energy sources have caused extensive concern. Distributed Generation (DGs) like wind turbine, photovoltaic (PV) can satisfy the expand energy demand, reduce the environment pollution and promote the coordinated development of environment and economy [1]. As penetration level of DG connected to utility grid increases, there are various issues needed to be considered such as islanding situation and power quality (PQ) problems [2]. Islanding operation forms because of fault or other reasons in power system when DG fails to detect power state and can not quickly cut off from public Grid and supply local loads instead [3]. This results to negative influence on public power system, such as the safety hazards to utility personnel, PQ disturbances and serious damages of equipment. Hence, it is important to detect islanding and PQ disturbances correctly and efficiently to increase the reliability of power supply [4].

Recently, several approaches have been proposed in detection islanding and classification of PQ disturbances which could be roughly divided into three groups: active, passive, and communication-based methods [5]. Generally, active methods analyze the changes of output signals by injecting a small disturbance into the system [6]. And the passive methods for effective detection of disturbances are based on observing system parameters and selecting appropriate thresholds. Active methods can bring about PQ problems, while passive methods are affected by non-detection areas and need to select the suitable threshold [7].

Nowadays, new methods have been proposed about wavelet transform (WT), artificial neural network (ANN), and the hybrid of the two [8]. The wavelet transform (WT) is useful in detecting PQ problems and analyzing signal in both time and frequency domains [9]. Artificial neural networks (ANNs) can be used to classify different PQ problems according to the results of WT [10].

In FNNs, the main object is to find the appropriate combination of connection weights and biases for purpose of achieving the least error in the process of learning. A wide variety of heuristic optimization methods have been adopted by training FNNs: Simulated Annealing (SA), Genetic Algorithms (GAs) [11], Particle Swarm Optimization (PSO) algorithms [12], and Differential Evolution (DE) [13]. In addition, SA and GA, suffering from low convergence speed, could avoid trapping in local minima. According to [14], in the aspect of avoiding the drawbacks of convergence speed, PSO is one of the effective and reliable training algorithms.

Based on the above analysis, this paper proposes an islanding detection and classification method based on Wavelet Transform (WT) and modified FNN in hybrid power system. Discrete wavelet transform (DWT) is used to decompose the negative sequence voltage signal at PCC into different frequency bands [15]. The energy content and SD of the wavelet details calculated will be fed to a trained FNN which will be able to predict the event by understanding the pattern of input feature vector. Apart from being efficient and accurate [15], the presented method makes no difference with the power quality and high classification accuracy compared to traditional techniques.

This thesis is organized as below: Grid-connected multiple DGs is presented in Section II and the proposed Wavelet Transform – PSOFNN based islanding detection technique is fully described in Section III. Then, the simulated results and discussions are described in Section IV. The conclusion is showed in Section VII.

2. Grid-connected multiple DGs

To describe the proposed method for islanding detection, a hybrid distribution (HS) system model is built by MATLAB. The HS designs a photovoltaic power plant with a rated power of 250 kW with constant light of 1000 W/m^2 and a wind power plant with a power rating of 1.5 MW with constant wind speed of 12 m/s. The photovoltaic array comprises of 86 parallel strings. Each of them consists of 7 Sun Power SPR-415E modules connected in series. A 3-level IGBT bridge based on PWM is used to control the converter. Wind turbine with power of 1.5MW and PV with power of 250kW connects to a 10 kV distribution system. Dome-fed induction generators (DFIG) wind turbines are made up of wound-rotor induction generators and an AC/DC/AC IGBT-based PWM converter [16]. A typical hybrid distribution system with two distributed generators is presented in Figure 1. The hybrid power system consists of PV and Wind Power plant connected to a 10 kV Grid at point of common coupling (PCC) [6]. PQ disturbances considered in this study are Line to Line (L-L), Line to Ground (L-G) fault, Line to Line to Ground fault and Non-Linear load switching.

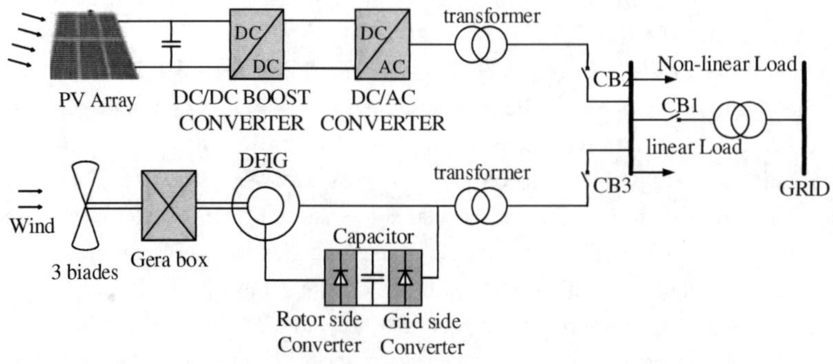

Figure 1. A hybrid power system with multiple DGs.

3. Detection and classification approach

In this chapter, the detection and classification method based on Wavelet Transform - PSOFNN is explained. The proposed method is on the basis of the analysis of transient process and identification of PQ events such as faults, switching and so on [15]. After the wavelet decomposition of the negative sequence voltage signal at the PCC [6], the eigenvalues of different scales are calculated and sent to a trained modified FNN model in order to classify different PQ disturbances.

3.1. Discrete Wavelet Transform (DWT)

WT decomposes data and functions in the time–frequency plane and is a choice for analyzing non-stationary signals emerged when island occurs [6].

In this paper, the negative sequence components of voltage signals at PCC are used as the input signals of the wavelet transform. When fault occurs, it can be noticed that the decomposition coefficients of the input signals by WT provide valuable information and represent the influence of any disturbances [17]. Daubechies4 (dB4) mother wavelet was proved to perform well in feature extraction for islanding detection [17].

Given a function v(t), the continuous wavelet transform (CWT) could be described as follow [6]:

$$CWT(v,x,y)=\frac{1}{\sqrt{a}}\int_{-\infty}^{+\infty} v(t)\varphi^{*}(\frac{t-y}{x})dt \quad (1)$$

Where x and y are scaling and translation constants, and φ is the wavelet function. Through the discrete wavelet transform, wavelet transform of sampling waveform could be gained:

$$DWT(v,x,y)=\frac{1}{\sqrt{x_0{}^{m}}}\sum_{k} v(k)\varphi^{*}(\frac{n-kx_0{}^{m}}{x_0{}^{m}}) \quad (2)$$

3.2. The Choice of features

To detect islanding and PQ disturbances, characteristic indicators like the energy content of the signal and Standard Deviation (SD) are extracting from negative sequence voltage by WT.

The Parseval's theorem states that the total energy of the signal in the time domain is equal to the total energy of the signal in the frequency domain, that is, the total energy of the signal after Fourier transform remains unchanged. Mathematical framework is shown as follows [18]:

$$E_{signal} = \frac{1}{T}\int_0^T |v(t)|^2 dt = \sum_{k=-\infty}^{\infty} |c(k)|^2 + \sum_{j=j_0}^{\infty}\sum_{k=-\infty}^{\infty} |d_j(k)|^2 \tag{3}$$

$$c_j(k) = \sum_{m=-\infty}^{\infty} c_{j+1}(m)h(m-2k) \tag{4}$$

$$d_j(k) = \sum_{m=-\infty}^{\infty} c_{j+1}(m)h_1(m-2k) \tag{5}$$

$$h_1(k) = (-1)^k h(1-k) \tag{6}$$

Where T represents the period of time, $c_j(k)$ and $d_j(k)$ are the quadrature mirror filters (QMF), $h(k)$ and $h_1(k)$ represent the scaling function and wavelet function.
The energy of signal and SD [19] for the detail coefficients is calculated in Equation (7) and Equation (8).

$$\|X\|_2 = (\sum_{i=1}^{N} |x_i|^2)^{1/2} \tag{7}$$

$$SD = \frac{1}{N-1}\sum_{i-1}^{N}(x_i - \mu)^2 \tag{8}$$

Where the amplitude of i_{th} sample of Wavelet Transform of sampling signal is x_i, N represents the number of samples and μ is the mean of the Wavelet Transform of signal.

3.3. Modified FNN: PSOFNN

Artificial neural networks (ANNs) had been successfully applied in many aspects in power systems operation and control. One of functions of ANN is often designed as classifiers because of the ability of learning, complex nonlinear mapping and parallel computing [20].
Artificial neural network is extensively interconnected by a large number of simple processing units (called neurons) [21] that solve specific problems in parallel. A simple neuron with input vector P is shown in Figure 2. The goal of Artificial Neural Networks (ANN) is that the network shows some desired behaviors by adjusting these parameters. Therefore, you can train the network to perform particular functions by changing the parameters.

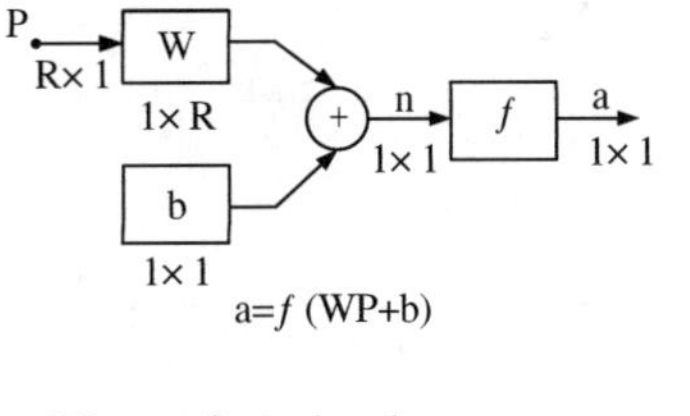

Figure 2. A simple neuron.

In this paper, PSO is applied to update the weights and biases of a network for the purpose of minimizing the mean square error (MSE) so as to the network will implement particular functions [21].
PSO is a heuristic optimization approach which is proposed by Kennedy [22], which searches for the target minimum by a number of particles in the space.
Every particle updates its position according to the current position and velocity in order to minimize the distance to global best position. It can be described as follows [23]:

$$v_i^{t+1} = wv_i^t + c_1 \times rand \times p(pbest_i - x_i^t) + c_2 \times rand \times (gbest - x_i^t) \quad (9)$$

$$x_i^{t+1} = x_i^t + v_i^{t+1} \quad (10)$$

Where v_i^t is the velocity of particle i, w is a weighting function, $c_j$is an acceleration coefficient, $rand$ is a random number between 0 and 1, $pbest_i$ is the pbest of agent i at iteration t, and gbest is the best solution.
The fitness function is introduced as follows [24]:

$$f(s_j) = (1 + \exp(-(\sum_{i=1}^{n} w_{ij}x_i - \theta_j)))^{-1}, j = 1, 2, ..., h \quad (11)$$

$$s_j = \sum_{i=1}^{n} w_{ij}x_i - \theta_j \quad (12)$$

$$o_{k=} \sum_{i=1}^{h} w_{kj} f(sj) - \theta_k, \quad k = 1, 2 \ldots, m \quad (13)$$

$$E_k = \sum_{i=1}^{m} (o_i^k - d_i^k)^2 \quad (14)$$

$$E = \sum_{k=1}^{q} \frac{E_k}{q} \quad (15)$$

$$fitness(X_i) = E(X_i) \quad (16)$$

Where n is the number of the input nodes, w_{ij} is the connection weight from the i_{th} node in the input layer to the j_{th} node in the hidden layer, $h_j$is the bias of the j_{th} hidden node.
A typical FNN with two layers is shown in Figure 3.

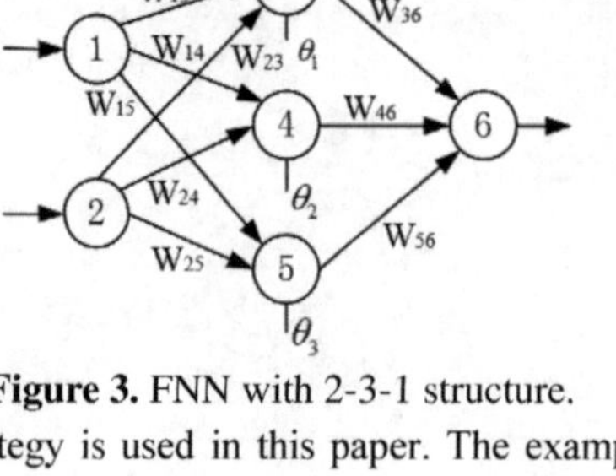

Figure 3. FNN with 2-3-1 structure.

Moreover, the matrix encoding strategy is used in this paper. The example of Figure 3 is shown as follows [24]:

$$particle(:,:,i) = \left[W_1, B_1, W_2^{'}, B_2 \right] \tag{17}$$

$$W_1 = \begin{bmatrix} w_{13} & w_{23} \\ w_{14} & w_{24} \\ w_{15} & w_{25} \end{bmatrix}, B_1 = \begin{bmatrix} \theta_1 \\ \theta_2 \\ \theta_3 \end{bmatrix}, W_2^{'} = \begin{bmatrix} w_{36} \\ w_{46} \\ w_{56} \end{bmatrix}, B_2 = [\theta_4] \tag{18}$$

Where W_1 and B_1 are respectively correspondent with weight matrix and bias matrix in the hidden layer, W_2' and B_2 are respectively correspondent with transpose of weight matrix and bias matrix in the output layer.

4. Simulink results and discussions

This section consists of the waveforms obtained on computer simulation of various events in MATLAB/Simulink. SD and the content of energy are calculated to train PSOANN. Simulation time is 1.5 seconds. Disturbances are introduced from 0.6 to 0.9 seconds.

4.1. Islanding Detection for Multiple DG systems.

Waveforms are shown in normal operation and islanding operation in Figure 4 and Figure 5.

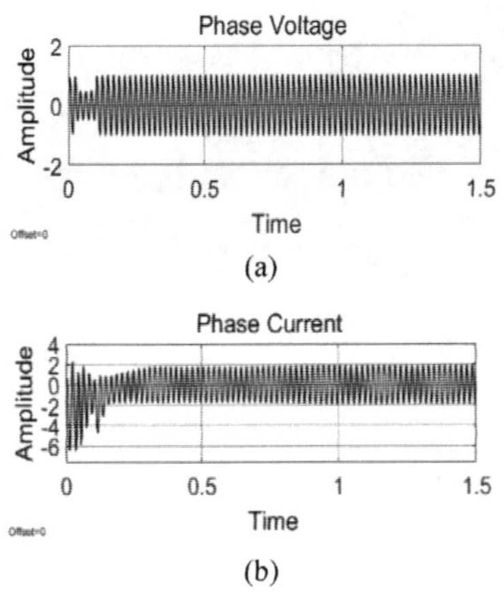

Figure 4. signals at normal operation at PCC. (a)Phase Voltage (b) Neg.Seq.Voltage (c)Phase Current

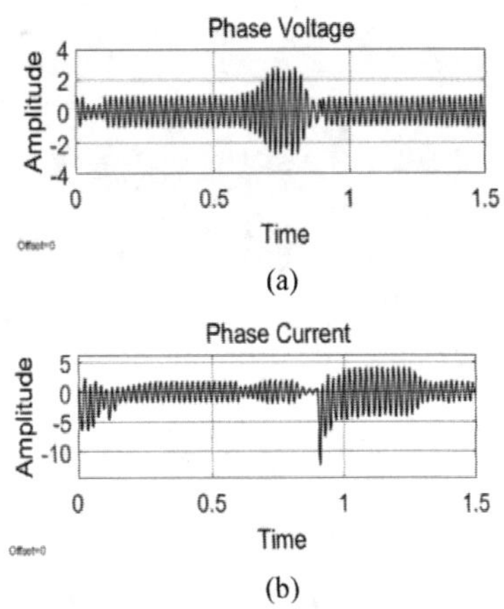

Figure 5. signals at islanding operation at PCC. (a)Phase Voltage (b)Neg.Seq.Voltage (c)Phase Current

4.2. PQ Disturbance Detection

IOP Conf. Series: Earth and Environmental Science **133** (2018) 012020 doi:10.1088/1755-1315/133/1/012020

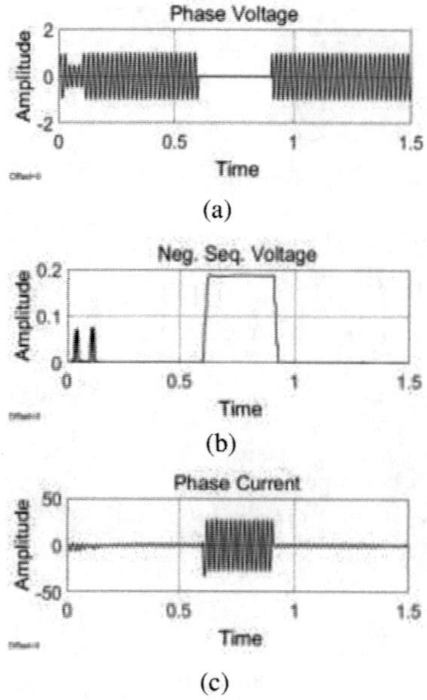

Figure 6. Signals at Line-Ground operation at PCC. (a)Phase Voltage (b)Neg.Seq.Voltage (c)Phase Current

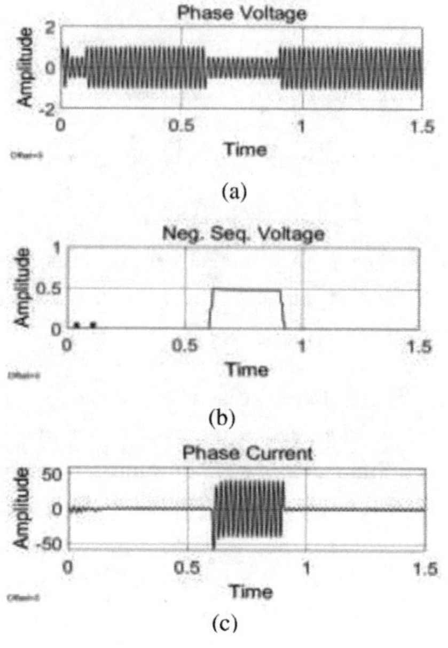

Figure 7 signals at Line-Line Fault operation at PCC. (a)Phase Voltage (b)Neg.Seq.Voltage (c)Phase Current

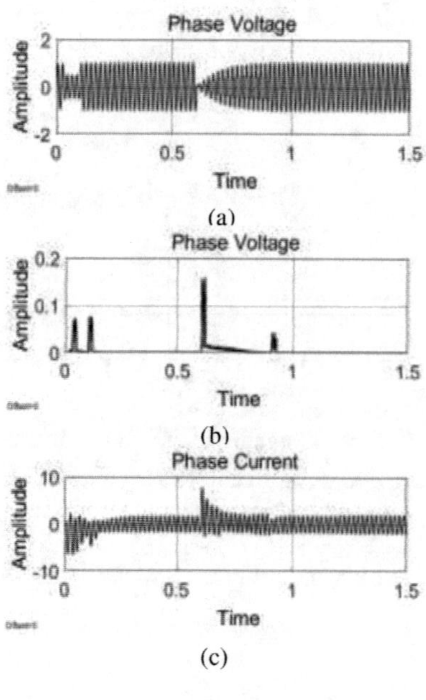

Figure 8. signals at Non-Linear Load Switching operation at PCC. (a)Phase Voltage (b)Neg.Seq.Voltage (c)Phase Current

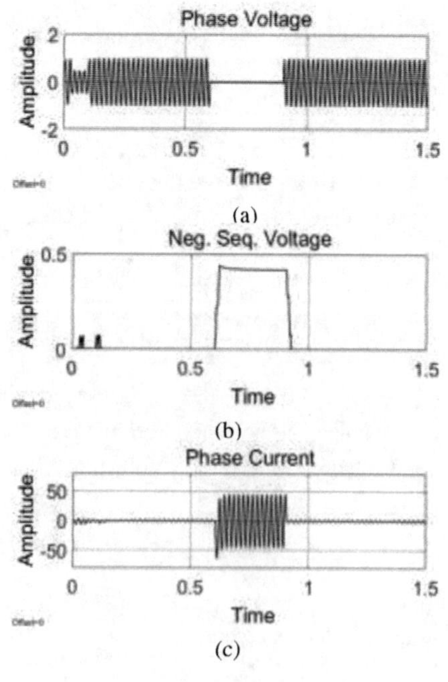

Figure 9. signals at Line-Line-Ground operation at PCC. (a)Phase Voltage (b)Neg.Seq.Voltage (c)Phase Current

IOP Conf. Series: Earth and Environmental Science **133** (2018) 012020 doi:10.1088/1755-1315/133/1/012020

It is evident from Figure 6 to Figure 9 that some differences exist in the negative sequence during different types of events can be noticed. These differences reflect on WT indices like SD and energy of the signal which are shown in Table 1 will be able to classify the events with high prediction accuracy. Meantime, THD changes abruptly during Islanding and it is hard to fix a threshold to detect PQ events. Compared to detection of harmonics method, new proposed approach performs well based on WT.

Table 1. Feature vectors for training.

Events	THD(%)			SD			Energy
Normal	0.94	9.6089e-5	5.2496e-5	5.5040e-5	3.5525e-5	2.5358e-5	0.0018
Islanding	20.46	0.0433	0.0535	0.0610	0.0209	0.0037	1.8193
L-L Fault	0.3	0.1248	0.0370	0.0210	0.0063	9.2025e-4	1.4265
L-G Fault	0.39	0.0482	0.0121	0.0068	0.0018	3.1991e-4	0.5065
LL-G Fault	0.17	0.1503	0.0215	0.0224	0.0078	9.1331e-4	1.0177
Non-Linear	15.79	0.0356	0.0271	0.0276	0.0076	0.0023	0.6294

4.3 Classification of PQ Disturbances based on Modified ANN

The data are divided two groups, one for training FNNs and the other for test the accuracy of classification. The example is randomly distributed and divided into training and test cases in the ratio of 70%: 30%. Figure 10 shows the convergence curves of PSOFNN based on averages of the MSE. The prediction result of the trained ANN is then compared with original label of the test example to get the accuracy. The total number of data is 150. Table 2 shows Labels for FNN Training with different events and 97.8% prediction accuracy is obtained.

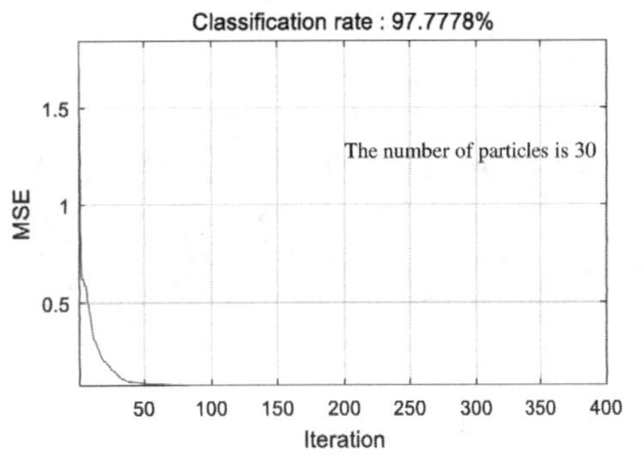

Figure 10. Convergence curves of PSOFNN.

Table 2. Labels and prediction results.

Events	Labels	Accuracy
Normal Operation	1	
Islanding	2	
Line-Line Fault	3	$\frac{44}{45} \times 100\% = 97.8\%$
Line-Ground Fault	4	
Line-Line-Ground	5	
Non-Linear load	6	

5. Conclusion and Future Work

This paper has given a presentation for islanding detection and classifying of various PQ problems in hybrid power system based on WT - PSOFNN method. Various situations such as islanding event, three-phase fault and PQ disturbances had been studied. PSO is employed to optimal the selection of the weights and biases of a network for purpose of minimizing the mean square error MSE. The evaluation index wavelet details are extracted from the negative sequence voltage signal at PCC by DWT and fed to a trained FNN that is able to distinguish between islanding and PQ problems. The Simulink results show that it can predict the type of the events with high prediction accuracy of 97.8%. The results of FNN validate the quality of WT indices taken and the effectiveness of proposed method in detection and classification of islanding and PQ disturbances in the hybrid power system.

Future work should focus on the following issues: 1) Testing more hybrid power system by proposed method. 2) Applying hybrid of PSO and other heuristic optimization methods to achieve higher convergence speed.

6. References

[1] Mirjalili S and Hashim S Z M 2010 *Int. Conf. on Computer and Information Application* pp 374-377

[2] Mohanty S R, Ray P K and Kishor N 2013 *J. International Journal of Electrical Power and Energy Systems* **44(1)** 764-777.

[3] Priyadharshini K M, Srinivasan S and Srinivasan C 2014 *IEEE Int. Conf. on Computational Intelligence and Computing Research* pp 1-6

[4] Samantaray S R, Pujhari T M, and Subudhi B D 2009 *IEEE International Conference on Power Systems (ICPS'09)* pp 1-6.

[5] PK Ray, SR Mohanty and N Kishor 2011 *Electric Power Systems Research* **81(3)** 805-819

[6] Puthenpurakel S P and Subadhra P R 2016 *Int. Conf. on Next Generation Intelligent Systems (ICNGIS)* pp 1-5

[7] Jang S I and Kim K H 2002 *Power Engineering Society Winter Meeting* vol 2 pp 800-806

[8] Ray P K, Kishor N and Mohanty S R 2012 *J. IEEE Transactions on Smart Grid* **3(3)** 1082-94

[9] Wang Y F and Ula AHMS 1991 *J. Electric Power Systems Research* **21(3)** 195-202

[10] Rahman M A 2006 *Int. Conf. on Electrical and Computer Engineering* pp 462-465

[11] Han X H, Xiong X, and Duan F 2015 *J. Applied Intelligence* **43(4)** 855-873

[12] Deep K and Bansal J 2008 *J. Nature Inspired Cooperative Strategies for Optimization (NICSO 2007)* **129** 453-460

[13] SR Mohanty, PK Ray, N Kishor and BK Panigrahi 2013 *International Journal of Electrical Power & Energy Systems* **44(1)** 764-777

[14] JF Chen, Do Q H and Hsieh H N 2015 *J. Algorithms* **8(2)** 292-308

[15] Fayyad Y and Osman A 2010 *In Electric Power and Energy Conference (EPEC)* pp 1-6

[16] Arun Bhaskar M, Magdal R and Dash S S 2012 *J. Engry Procedia* **14** 1536-41

[17] Puthenpurakel S P and Subadhra P R 2016 *Int. Conf. on Next Generation Intelligent Systems (ICNGIS)* pp 1-5

[18] Bala B, Babu S and Devi N R 2012 *J. International Journal of Engineering Science and Advanced Technology* **2(5)** 1040-46

[19] Feng, P, Wang X, Song J, Song Y, et al. 2014 *IEEE 12th Int. Conf. on Dependable Autonomic and Secure Computing(DASC)* pp 532-535.

[20] Ray P K, Mohanty S R and Kishor N 2011 *J. Electric Power Systems Research* **81(3)** 805-819

[21] Mohammadi M, Danandeh A, Ojaroudi N, et al. 2015 *J. Complexity* **21(2)** 309-324

[22] Tarafdar Hagh, Mehrdad, Homayoun Ebrahimian and Noradin Ghadimi 2015 *Frontiers in Energy* **9(1)** 75-90

[23] J. Kennedy, R.C. Eberhart 1995 *Proceedings of IEEE Int. Conf. on Neural Networks* **4** 1942-48

[24] Demertzis K and Iliadis L 2015 *International Symposium on Statistical Learning and Data Sciences* (Springer, Cham) pp 223-233

IOP Conf. Series: Earth and Environmental Science **133** (2018) 012021 doi:10.1088/1755-1315/133/1/012021

Study on the microwave catalytic pyrolysis characteristics and energy consumption analysis of oil shale

Chen Chunxiang[1,3], Cheng Zheng[1,3], Xu Qing[2], QIN Songheng[1,3]

[1] School of Electrical Engineering, Guangxi University, Nanning 530004, China;

[2] School of Mechanical and Power Engineering, Guang Ocean University, Zhanjiang 524000, China

[3] National Demonstration Center for Experimental Electrical Engineering Education（Guangxi University), Nanning 530004, China

Abstract. In order to explore the high-efficient utilization of oil shale, the effects of different microwave powers and different kinds of catalysts (metal oxides and metal salts) on the temperature characteristics and product yield towards the oil shale are investigated by microwave catalytic pyrolysis. The results show that the effect of microwave power on the heating and pyrolysis rates of oil shale is significant, and the maximum output of shale oil is 5.1% when the microwave power is 1500W; CaO has a certain effect on the temperature rise of oil shale, and MgO and CuO have a certain degree of inhibition, but the addition of three kinds of metal oxidation is beneficial to increase the shale oil production; From the perspective of unit power consumption and gas production, the catalytic effect order of three kinds of metal oxides is MgO> CaO> CuO; The addition of three kinds of metal salts is favorable for the increase of pyrolysis temperature of oil shale, after adding 5% $ZnCl_2$, the unit power consumption of shale oil and pyrolysis gas increases by 62.60% and 81.96% respectively. After adding 5% NaH_2PO_3, the unit power consumption of shale oil increases by 64.64%, and reduces by 9.56% by adding 5% $MgCl_2$.

1. Introduction

With the increasing tension in oil and gas resources and the rising of international oil prices, unconventional energy development and utilization have also been taken seriously. Oil shale is recognized as having the potential for alternative fossil fuels because of its rich reserves, ease development and utilization [1]. Oil shale is a high-ash sedimentary rock containing combustible organic matter, which belongs to unconventional energy and oil content of which is generally higher than fossil fuels by approximately 3.5% [2].

At present, the utilization of oil shale is mainly adopting the traditional heating method-----dry distillation technology, but this way exists the shortcomings of long-time consuming and low oil

IOP Conf. Series: Earth and Environmental Science **133** (2018) 012021 doi:10.1088/1755-1315/133/1/012021

production[3]. Microwave pyrolysis is a kind of new heating method, compared with conventional pyrolysis technology, this new method has the advantages of faster pyrolysis speed, higher efficiency, more uniform pyrolysis and easier control[4]. Therefore, this technology received great attention by many scholars for oil shale pyrolysis.

In this paper, the microwave reactors were used to study the effects of different microwave powers, metal oxides and metal salts on the pyrolysis characteristics of oil shale and the yield of three-phase products. And the absolute productivity of oil shale microwave pyrolysis was studied by combining the energy consumption and energy output evaluation method. The results can provide theoretical guidance for the exploitation of oil shale by microwave.

2. Experimental facilities and methodology

2.1. Sample preparation

The oil shale used in the experiment is from the kiln and the basin in Gansu. The industrial analysis and elemental analysis of oil shale are shown in Table 1.The experimental microwave absorbers were metal oxides (CaO, CuO and MgO) and metal salts ($MgCl_2$, $ZnCl_2$ and NaH_2PO_3). Before the start of the experiment, the activated carbon is dried at 105°C for 24 hours thermostatically. For coherent comparison, the reactants (dry basis) and the microwave absorptions used for each experiment were 30 g and 5%, respectively.

Table 1. Elemental analysis, proximate analysis and low calorific value of oil shale

Sample	Elemental analysis (wt %)					Industry analysis (wt %)			Low calorific value (MJ/kg)
	C	H	O	N	S	V	A	FC	$Q_{net,ar}$
Oil shale	15.68	2.3	-----	0.57	0.63	26.97	72.13	0.90	5.56

2.2. Experimental facilities

The experimental device of microwave system is shown in Figure 1. Experiment device mainly consists of microwave generation and control system, temperature measurement system, material reaction system, pyrolysis products collection system and data acquisition system.

Figure 1. The reactor of experimental device

3. Results and discussion

3.1. Effect of Different Microwave Power on of Oil Shale Pyrolysis

The sample weight for each pyrolysis experiment was 30 g, and three microwave powers were selected (750 W, 1500 W and 2250 W). The curves for temperature vs pyrolysis time of the oil shale at different microwave powers measured by thermocouple is shown in Fig. 2 (a). The percentage distributions of the oil shale pyrolysis products at different microwave powers is shown in Fig. 2 (b).

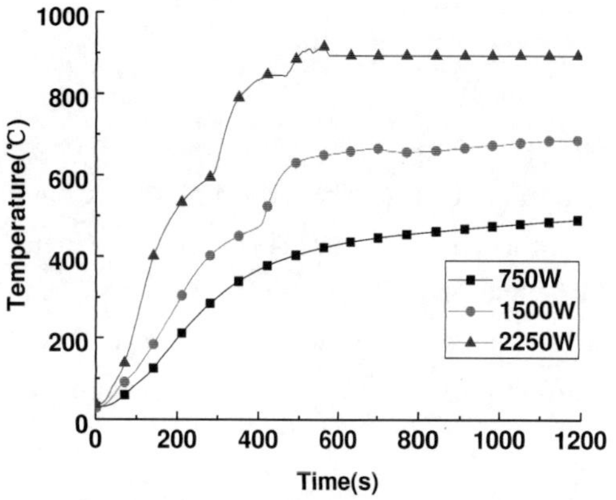

Figure 2(a). The impact of different microwave powers different on the temperature rise

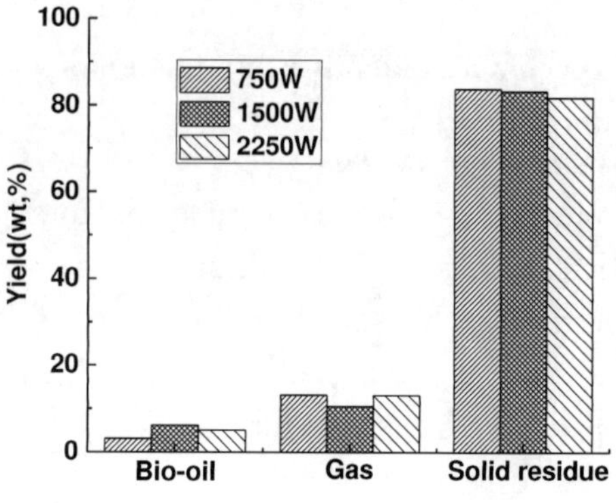

Figure 2(b). Percentage of product at microwave power

It can be seen from Fig. 2 (a) that the oil shale temperature increased gradually when the pyrolysis time prolonged. At the beginning, the oil shale was heated up quickly, when the temperature reached its maximum there was no significant change observed. This could be ascribed that the amount of heat generated by the pyrolysis gas was close to the heat (generated by the microwave) absorbed by material [5].With the increase of microwave power, the material reached the maximum temperature with less time, and the final temperature of the material was also higher, which may be due to the greater microwave power, which made the microwave field interaction among water molecules become faster, and it accelerated the pyrolysis of the material[6].

As can be seen from Table 2, with the increase of microwave power, the average heating rate and the maximum heating rate were increased. This is mainly because when the microwave power is small, the oil shale involved in pyrolysis is less, resulting in a small heating rate. And with the increasing microwave power, microwave energy absorbed by oil shale increases, the intensity of molecular motion also increased[7]. And the impact of microwave power on the heating rate is very obvious, indicating that by controlling the microwave power can be very good to achieve the purpose of controlling the temperature in practice.

As can be seen from Fig. 2 (b), when the microwave power is 1500W, the oil production reaches its maximum value of 5.1%; when the microwave power is 750W, the amount of residue becomes the largest, reaching 83.73%. With the increase of microwave, the amount of gas produced decreases first and then increases, and the oil production increases first and then decreases. When the microwave power is 1500W, due to the larger power, oil shale pyrolysis is nearly more completed, resulting in a large number of shale oil and pyrolysis gas. While the microwave power of 2250W, the polymerization reaction and cracking reaction occurred in the material competed with each other[8], during this period, the cracking reaction rate is greater than the polymerization reaction rate, t severe secondary decomposition reaction occurred, shale oil production rate is smaller than the shale oil decomposition rate, making the oil production decreased.

Table 2. The impact of different microwave powers on the maximum and average heating rates

Microwave power	Average heating rate (°C/s)	Maximum heating rate (°C/s)
750W	0.378	1.41
1500W	0.538	4.031
2250W	0.703	5.031

In summary, 1500W is the best power for oil shale pyrolysis, its highest oil production is 5.1%, so in the experiment 1500W is selected as the pyrolysis power.

3.2. Effect of Different Metal Oxide on Pyrolysis of Oil Shale

The microwave power, the reactants (dry basis) and the microwave absorption (CuO、MgO and CaO) used for each experiment were 1500W, 30g and 5%, respectively. The pyrolysis temperature curves of the oil shale measured by thermocouple is shown in Fig. 3 (a). The percentage distribution of the oil shale pyrolysis products at different microwave power is shown in Fig. 3 (b).

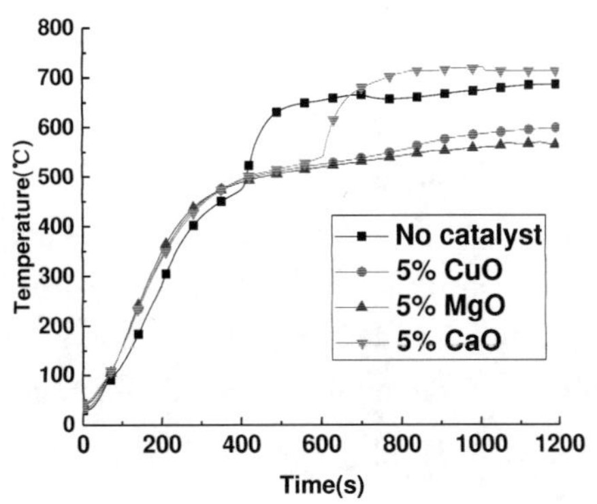

Figure 3(a). The impact of different metal oxides different on the temperature rise

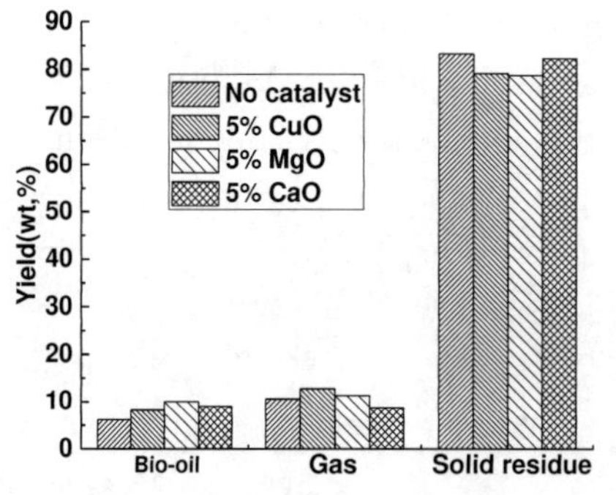

Figure 3(b). Percentage of product under metal oxides

It can be seen from Fig. 3 (a) that the curves of the samples with MgO and CuO were very close to each other. After 413s, the temperature was always lower than pure oil shale, and the final pyrolysis temperatures of which were lower than the pure oil shale by 132.8°Cand 88.8°C, respectively. This could be resulted from that MgO and CuO have no significant effect on the chemical reaction within the material and it increased the mass transfer and heat transfer during pyrolysis. After adding calcium oxide, the temperature rise of the sample is very close to that of pure oil shale, and the final temperature is only 26.4°C higher than that of pure oil shale. Because CaO is a weak-absorbing material that absorbs a small amount of microwave while increasing the heat transfer resistance of the specimen.

It can be seen from Fig. 3 (b) that the amount of residue after adding CuO, MgO and CaO were 4.16%, 4.53% and 1.00%, respectively, which were lower than that of pure oil shale, and the shale oil production with these three additions were 2.00%, 3.80% and 2.83%, respectively, which were higher than pure oil shale. Studies have shown that [7] CuO, MgO and CaO and other alkaline earth metal oxides have the catalytic effect, contributing to shale oil pyrolysis, which may be the reason for the decline in residue. The gas production of the sample adding CuO and MgO were 2.16% and 0.73% higher than that of the pure oil shale, but which was reduced by 1.84% when adding CaO. This may be due to the reaction of CO_2 and CaO to produce $CaCO_3$, resulting in a decrease in gas production [5].

From the unit power consumption of the three-phase product shown in Fig. 3 (c), it can be seen that the unit power consumption of shale oil and pyrolysis gas increases with the addition of metal oxide,The unit power consumption of the pyrolysis residue of the samples with MgO and CaO increased and the unit power consumption of the pyrolysis residue of the CuO sample was decreased.

From the perspective of unit power consumption and gas production, it can be seen that the catalytic effect of the three kinds of metal oxides is MgO> CaO> CuO.

IOP Conf. Series: Earth and Environmental Science **133** (2018) 012021 doi:10.1088/1755-1315/133/1/012021

Table 3 . The impact of different metal oxides on the power consumption

sample	Power (W)	Time (h)	Power consumption (kWh)
No catalyst	1500	0.304	0.681
CuO	1500	0.311	0.697
MgO	1500	0.253	0.566
CaO	1500	0.234	0.524

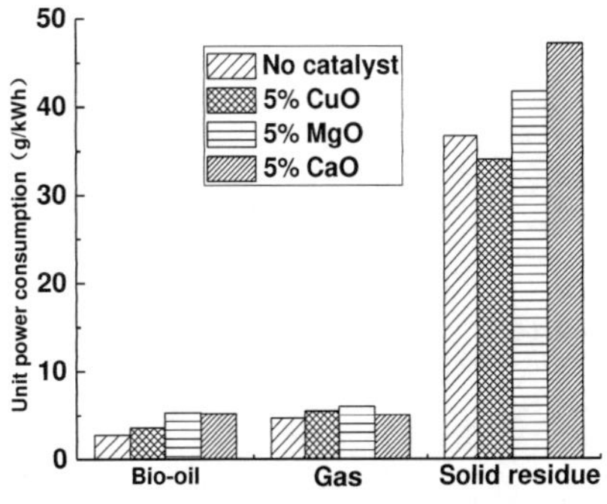

Figure 3(c). The impact of different metal oxides on the mass of products of unit power consumption

3.3. Effect of Different Metal Salt on Pyrolysis of Oil Shale

The microwave power, reactants (dry basis) and the microwave absorption（$MgCl_2$、$ZnCl_2$ and NaH_2PO_3）used for each experiment were 1500W, 30 g and 5%, respectively. The pyrolysis temperature curves of the oil shale measured by thermocouple is shown in Fig. 4 (a). The percentage distribution of the oil shale pyrolysis products at different microwave power is shown in Fig. 4 (b).

It can be seen from Fig. 4 (a) that the reaction rate of the material was accelerated after adding three kinds of metal salts, and the final temperatures of $MgCl_2$ and NaH_2PO_3 were 29.1% and 26.6% higher than that of pure oil shale respectively. The four temperature rising curves have a rapid warming range, and the reason for the rapid warming range is that the residue formed after pyrolysis of oil shale presents good microwave-absorbing ability [9], which can promote the pyrolysis reaction of the sample. After adding the three metal salts, the time to enter the rapid warming range is in advance, which may be resulted from the addition of the metal salt so that the time of the residue formation in advance.

Fig.4 (b) shows that the three-phase production percentages of sample with addition of $MgCl_2$, $ZnCl_2$ and NaH_2PO_3, the amount of residue decreased by 5.28%, 2.08% and 1.88%. This suggests that the addition of metal salts promote the pyrolysis of oil shale. The shale oil production of samples with $MgCl_2$ and $ZnCl_2$ decreased by 1.53% and 3.07%, while the gas production increased by 6.76% and 1.76% respectively. This is because of the addition of $MgCl_2$ and $ZnCl_2$ that promote the secondary pyrolysis of shale oil, resulting in an increase in pyrolysis gas production and a decrease in shale oil production[10] . After adding NaH_2PO_3, shale oil production increased by 2.97% and pyrolysis gas decreased by 1.14%.This shows that NaH_2PO_3 is conducive to the production of shale oil.

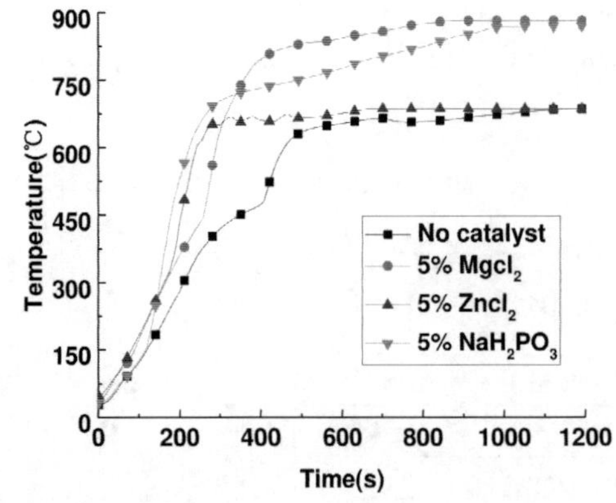

Figure 4(a). The impact of different metal salts on the temperature rise

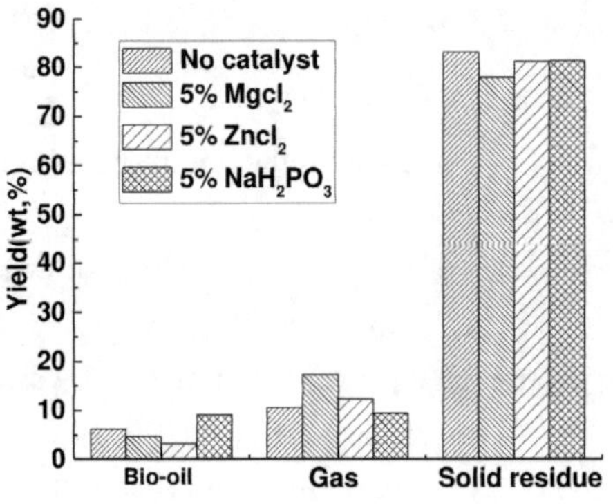

Figure 4(b). Percentage of product under different metal salts

Table 4. The impact of different metal salts on the power consumption

sample	Power (W)	Time (h)	Power consumption (kWh)
No catalyst	1500	0.304	0.681
$MgCl_2$	1500	0.253	0.566
$ZnCl_2$	1500	0.195	0.437
NaH_2PO_3	1500	0.273	0.611

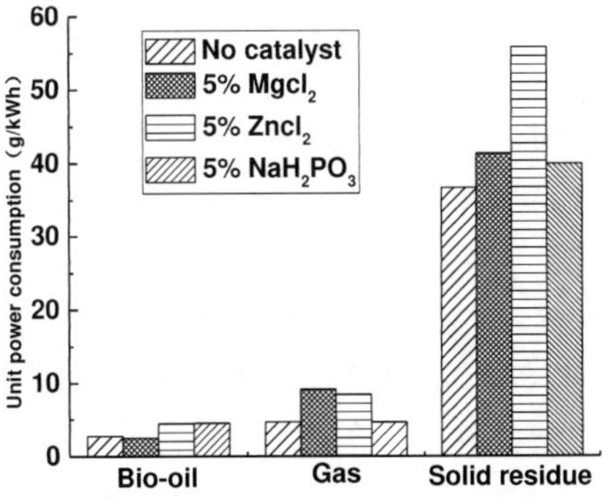

Figure 4(c). The impact of different metal salts on the mass of products of unit power consumption

It can be seen from Table 4 that after adding $MgCl_2$, $ZnCl_2$ and NaH_2PO_3, the electricity consumption decreased by 16.89%, 35.83% and 10.28%, respectively. It can be seen from Fig. 4 (c) that the addition of three kinds of metal salts can significantly improve the unit power consumption of pyrolysis residue and pyrolysis gas. After adding $ZnCl_2$ and NaH_2PO_3, the unit power consumption of shale oil increased and the unit power consumption of $MgCl_2$ sample was decreased. The unit power consumption analysis shows that the catalytic effect order of three kinds of metal salts was $ZnCl_2$> $MgCl_2$> NaH_2PO_3.

Compared with the addition of metal oxide and metal salt in 3.2 and 3.3, the catalytic effect of metal salt is generally better than that of metal oxide. The best catalytic effect is $MgCl_2$.

4. Conclusion

The greater the microwave power, the higher the average heating rate and the pyrolysis temperature of the oil shale. When the pyrolysis power is 1500W, the shale oil production reaches the maximum value of 5.1%.

After adding 5% CaO, the final temperature of the sample increased slightly, while with the addition of MgO and CuO, the pyrolysis temperature decreased by 12.69% and 17.73%, respectively.

The addition of three kinds of metal salts was beneficial to the increase of pyrolysis temperature of oil shale. After adding 5% $ZnCl_2$, the unit power consumption of shale oil and pyrolysis gas production increased by 62.60% and 81.96% respectively. After adding 5% NaH_2PO_3, the unit power consumption of shale oil increased by 64.64%, while decreased by 9.56% by adding 5% $MgCl_2$.

The catalytic effect of metal salts is generally better than that of metal oxides, and the $MgCl_2$ has the best catalytic effect.

References

[1] Al-Harahsheh A, Al-Ayed O, Al-Harahsheh M, et al. Heating rate effect on fractional yield and composition of oil retorted from El-lajjun oil shale[J]. Journal of Analytical & Applied Pyrolysis, 2010, 89(2):239-243.

[2] YAN Junwei, JIANG Xiumin, HAN Xiangxin,et al. A TG-FTIR investigation to the catalytic effect of mineral matrix in oil;shale on the pyrolysis and combustion of kerogen[J]. Fuel, 2013, 104(2):307-317.

[3] LO S L, HUANG Y F, CHIUEH P T, et al. Microwave Pyrolysis of Lignocellulosic Biomass[J]. Energy Procedia, 2017, 105:41-46.

[4] HUANG Y F, CHIUEH P T, LO S L. A review on microwave pyrolysis of lignocellulosic biomass[J]. Sustainable Environment Research, 2016, 26(3):103-109.

[5] Allawzi Mamdouh, Al-Otoom Awni, Allaboun Hussein, et al. CO_2 supercritical fluid extraction of Jordanian oil shale utilizing different co-solvents [J]. Fuel Process Technol,2011, 92 (10): 2016–2023.

[6] HU Z, MA X, CHEN C. A study on experimental characteristic of microwave-assisted pyrolysis of microalgae.[J]. Bioresource Technology, 2012, 107(3):487-493.

[7] Mokhlisse A, Chanaa M B, Outzourhit A. Pyrolysis of the Moroccan (Tarfaya) oil shales under microwave irradiation[J]. *Fuel*, 2000, 79(7): 733-742.

[8] LAN X, LUO W, SONG Y, et al. Effect of Temperature on the Characteristics of Retorting Products Obtained by Yaojie Oil Shale Pyrolysis[J]. Energy & Fuels, 2015,29 (12):7800–7806.

[9] Haykiri-Acma H, Yaman S, Kucukbayrak S. Effect of heating rate on the pyrolysis yields of rapeseed[J]. *Renewable Energy*, 2006, 31(6): 803-810.

[10] LEE J Y, CHAN Y, JUN S Y, et al. Comparison of several methods for effective lipid extraction from microalgae[J]. Bioresource Technology, 2010, 101(1):S75-S77.

IOP Conf. Series: Earth and Environmental Science **133** (2018) 012022 doi:10.1088/1755-1315/133/1/012022

Briquette fuel production from wastewater sludge of beer industry and biodiesel production wastes

P Nusong and S Puajindanetr

Department of Industrial Engineering, Chulalongkorn University, Bangkok, 10300, Thailand

Email: parinrat.ns@gmail.com

Abstract. The production of industrial wastes is increasing each year. Current methods of waste disposal are severely impacting the environment. Utilization of industrial wastes as an alternative material for fuel is gaining interest due to its environmental friendliness. Thus, the objective of this research was to study the optimum condition for fuel briquettes produced from wastewater sludge of the beer industry and biodiesel production wastes. This research is divided into two parts. Part I will study the effects of carbonization of brewery wastewater sludge for high fixed carbon. Part II will study the ratio between brewery wastewater sludge and bleaching earth for its high heating value. The results show that the maximum fixed carbon of 10.01% by weight was obtained at a temperature of 350 °C for 30 minutes. The appropriate ratio of brewery wastewater sludge and bleaching earth by weight was 95:5. This condition provided the highest heating value of approximately 3548.10 kcal/kg.

1. Introduction

The consumption of beverage products in Thailand is increasing in quality significantly every year, one of the beverage products is beer. According to the bank of Thailand statistics, beer's sale expanded 24.36% from 2004 to 2014 [1]. Following the sale expansion, it caused the environmental problems, especially wastewater pollution. During the production, a liter of beer, wastewater generates around 3-10 liters [2]. The most famous industrial wastewater treatment methods are biological treatment with activated sludge process. Therefore, a large amount of brewery wastewater sludges was generated, which required proper management before disposal. Most of these brewery wastewater sludges are disposed of in landfills that is the resulting in environmental impacts on surface and groundwater. Brewery wastewater sludge can be served as a potential renewable fuel, due to a significant amount of organic matter of adequately high heating value.

The oil and gas situation in Thailand drives the economy and increased sharply. As a result, the Thai government has been promoting domestic production biofuel as a substitute crude oil import, such as gasohol, palm diesel, and biodiesel. Mostly, the commercial scale biodiesel plants use crude palm oil as a raw material. The palm oil bleaching process as the most important activity in palm oil refinery, which removes contaminants and color [3]. Bleaching earth is one of the residuals in palm oil refinery that required proper management before disposal as same as brewery wastewater sludge.

Biodiesel production process is a chemical reaction called esterification between vegetable oils or animal fats and alcohol. The products of this reaction are biodiesel as a main product and glycerol as a by-product, which produces 100 kilograms of biodiesel, generates 11 kilograms as crude glycerol [4]. Crude glycerol from biodiesel production plant is not commonly used in the pharmaceutical and cosmetic industries because of its low purification, Thus, the aim of this study was to investigate the potential of the briquettes made from brewery wastewater sludge mixed with bleaching earth and crude glycerol as alternative fuel and wastes utilization.

2. Experimental methods

2.1. Materials

The technique of full factorial design was used to study the optimum condition of two factors, namely, temperature and time. Each experiment is performed by 3 replicated at 95% confidence interval. The carbonization process was operated at different temperatures (300, 350 and 400 ^{o}C) and times (30, 60 and 90 minutes) in a limited-air electrical furnace. During the briquetting process, certain amounts of bleaching earth were mixed with brewery wastewater sludge. Note that the type of briquetting process carried out here was a cold process using crude glycerol (30% by weight of the mixture) as the binder. Five different ratios (brewery wastewater sludge to bleaching earth) used in the briquetting process were: 55:45, 65:35, 75:25, 85:15 and 95:5. This experiment was observed by using mixture design. Each experiment is performed by 2 replicated at 95% confidence interval.

2.2. Testing procedures

The fixed carbon was determined after carbonization process at various temperature and time according to ASTM D3172-07 [5]. After the briquetting process, fuel briquettes were obtained and their heating values were measured based on ASTM D1989-97 [6].

3. Experimental results

3.1. Characteristics of Raw Materials

From table 1, it was found that brewery wastewater sludge exhibited the higher fixed carbon and heating value than bleaching earth.

Table 1. Properties of brewery wastewater sludge, bleaching earth and crude glycerol

Material	Fixed carbon (% wt)	Heating value (kcal/kg)
Brewery wastewater sludge	4.36	3,045.35
Bleaching earth	2.66	2,553.25
Crude glycerol	-	3,534.98

3.2. Optimum conditions for carbonization process

As shown in figure 1, with the temperature and time rising from 300 to 400 °C and 30 to 90 minutes, respectively, the fixed carbon of brewery wastewater sludge increases up to a certain point then decreases. It was found that increasing temperature and time caused the mass conversion [7] and devolatilization of organic compounds or hydrocarbons during carbonization [8]. But increasing high temperature affects the amorphous carbon atoms inside brewery wastewater sludge and removes surface functional groups [9].

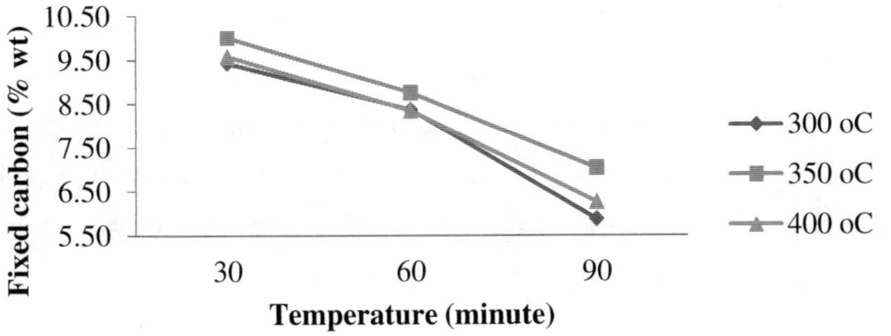

Figure 1. The effect of temperature and time for fixed carbon of brewery wastewater sludge.

3.3. Optimum ratios for briquetting process

The effect of ratios for briquette process is shown in figure 2. The increasing in an amount of brewery wastewater sludge and decreasing in an amount of bleaching earth ratio in briquette mixture, the heating value showed an incremental from 3,269.80 kcal/kg to 3,548.10 kcal/kg. It is because brewery wastewater sludge has a higher heating value than bleaching earth.

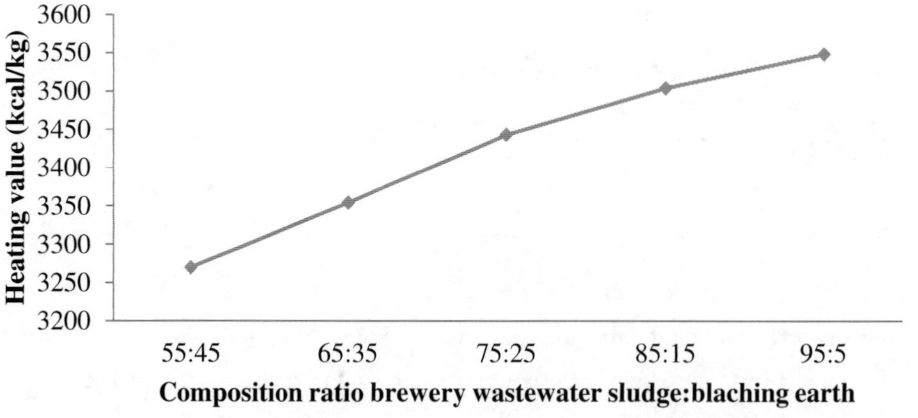

Figure 2. The effect of ratio for heating value of briquette.

3.4. Statistical analysis

It is evident from table 2 and table 3 that the fixed carbon and the heating value are affected by both factors. Firstly, analyzing the effect of each factors. It was found that both the temperature ($P < 0.05$) and time ($P < 0.05$) significantly affected the fixed carbon in brewery wastewater sludge. Also, the ratio of brewery wastewater sludge ($P < 0.05$) and bleaching earth ($P < 0.05$) significantly affected the heating value in briquette. Secondly, there was no significant interaction effect of temperature and time ($P > 0.05$). On the other hand, there was significant interaction effect of brewery wastewater sludge and bleaching earth ratio ($P < 0.05$).

Table 2. ANOVA table for the factorial design (fixed carbon is the output)

Term	DF	F-value	P-value
Temperature	2	17.06	0.000
Time	2	346.24	0.000
Temperature*Time	4	1.68	0.198

Table 3. ANOVA table for the mixture design (heating value is the output)

Term	Coef	SE Coef	P-value
Brewery Sludge	3567.5	6.951	> 0.000
Bleaching Earth	2476.2	70.246	> 0.000
Brewery Sludge*Bleaching Earth	770.5	127.728	0.001

3.5. Relationship between the heating value and the material ratio

The relationship between the heating value and the brewery wastewater sludge and bleaching earth ratio could be written in the equation 1:

$$HV_{CAL} = 0.7693m_{BS}HV_{BS} + 0.7693m_{BE}HV_{BE} + 0.7693m_{G}HV_{G} \quad (1)$$

where HV_{CAL} is the calculated heating value of the briquette (kcal/kg), 0.7693 is the conversion factor from 130% to 100%, m_{BS} is the mass fraction of brewery wastewater sludge (%), HV_{BS} is the heating value of brewery wastewater sludge after carbonization (kcal/kg), m_{BE} is the mass fraction of bleaching earth (%), HV_{BE} is the heating value of bleaching earth, m_G is the mass fraction of crude glycerol (%) and HV_G is the heating value of crude glycerol. The mass fraction of crude glycerol was 30% and the heating value of brewery wastewater sludge after carbonization, bleaching earth and crude glycerol were 3,480.65 kcal/kg, 2,553.25 kcal/kg and 3,534.98 kcal/kg, respectively.

Based on equation 1, the calculated heating value of the briquettes were obtained. Figure 3 shown the calculated heating value of the briquettes compared to the actual heating value. It was found that the actual heating value was correlated well with the calculated heating value within ±5% errors. It could be seen that increasing an amount of brewery wastewater sludge and decreasing in an amount of bleaching earth ratio in mixture led to increasing the heating value of briquette. The optimum brewery wastewater sludge and bleaching earth ratio found in this study was 95% by weight.

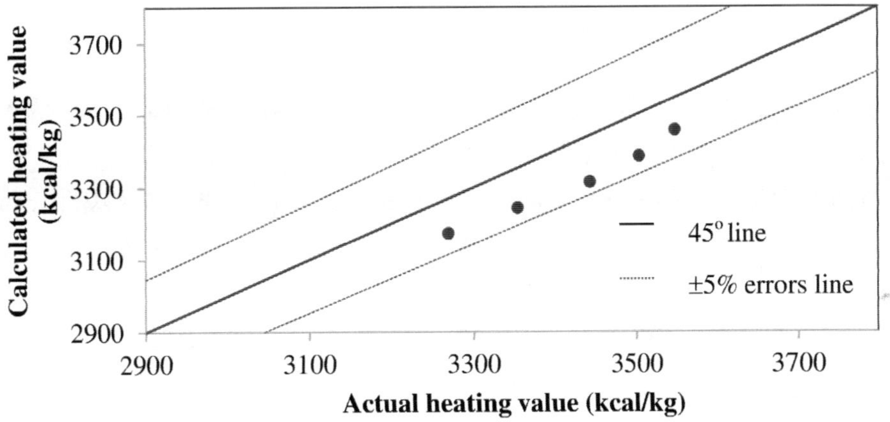

Figure 3. The comparisons of the calculated heating value and the actual heating value in difference ratio of material.

4. Conclusions

This research investigated the optimum condition for fuel briquettes produced from brewery wastewater sludge bleaching earth and crude glycerol. The results showed that the optimal condition for carbonization process was at temperature 350 °C for 30 minutes of heating time, which is provided highest fixed carbon. In addition, the highest heating value from this studied was from 95% by weight of brewery wastewater sludge mixed with bleaching earth as optimal ratio. To make the briquette made from brewery wastewater sludge mixed with biodiesel production wastes more efficient, the development such as compressive strength, the ratio of crude glycerol, reducing of ash and toxic gas emission are required in the future study.

References

[1] Bank of Thailand 2017 Private Consumption Indicators. http://www2.bot.or.th/statistics/ReportPage.aspx?reportID=108.

[2] Dorman L I 1975 *Variations of Galactic Cosmic Rays* (Moscow: Moscow State University Press) p 103

[2] Kanagachandran K and Jayaratne R 2006 Utilization potential of brewery waste water sludge as an organic fertilizer. *J. Institute of Brewing* **112**(2) 92-96.

[3] Simone M. S., Klicia A. S., Roberta C., Roland V., Christian S.,Wim D G and Antonio J A M 2014 Effect of type of bleaching earth on the final color of refined palm oil. *LWT - Food Science and Technology* 59(2) 1258-1264.

[4] Behr A, Eilting J, Irawadi K, Leschinski J and Lindner F 2008 Improved utilisation of renewable resources: new important derivatives of glycerol. *J. Green Chemistry* **10** 13–30.

[5] ASTM 1997 Standard test method for gross calorific value of coal and coke by microprocessor controlled isoperibol calorimeters. *Amarican Society for Testing and Materials.* ASTM D1989-97.

[6] ASTM 2007 Standard practice for proximate analysis of coal and coke. *Amarican Society for Testing and Materials.* ASTM D3172-07.

[7] Shrikalaa K., Yvan G and G.S. Vijaya R 2017 Optimization and characterization of hydrochar produced from microwave hydrothermal carbonization of fish waste. *J. Waste Management* **65** 159-168.

[8] Beiping Z., Sijiang X., Bo X., Dongke Y and Xiaoyuan J 2011 Mechanism of wet sewage sludge pyrolysis in a tubular furnace. *Int. J. Hydrogen Energy* **36**(1) 355-363

[9] Minori U., Lynda H. W., K. Thomas K., Chanel A F and Isabel M L 2011 Influence of pyrolysis temperature on biochar property and function as a heavy metal sorbent in soil. *J. agricultural and food chemistry* **59** 2501-2510

IOP Conf. Series: Earth and Environmental Science **133** (2018) 012023 doi:10.1088/1755-1315/133/1/012023

Effects of catalysts on combustion characteristics and kinetics of coal-char blends

Yingjie Hu, Zhiqiang Wang*, Xingxing Cheng, Ming Liu and Chunyuan Ma
National Engineering Laboratory for Coal-fired Pollutants Emission Reduction, Shandong University, 17923 Jingshi Road, Jinan 250061, PR China

Email: jackywzq@sdu.edu.cn

Abstract. The effects of Fe_2O_3, CaO, and MnO_2 on the combustion characteristics and kinetics of coal-char blends were investigated using thermogravimetric analysis (TGA) and differential scanning calorimetry (DSC). The results indicated that catalysts exhibited positive effects on the combustion characteristics of coal-char blends, especially in the initial period of coal-char blends combustion. With catalysts addition (mass 1.5%), it could improves volatile matter release, and reduces ignition point, promotes char to begin burning under lower temperature. The ignition index (C) was increased, respectively, by 27% for Fe_2O_3, 6% for CaO, 11.3% for MnO_2, and the combustion characteristic index (***S***) was increased respectively, by 29% for Fe_2O_3, 5% for CaO, 8.3% for MnO_2. In addition, two kinetic models (R_2 and F_1) were adopted to calculate the kinetic parameters in different stage of combustion processes. The results showed that with Fe_2O_3 or CaO addition, the activation energy at second stage decreases from 86.0 KJ/mol to 76.92 KJ/mol and 75.12 KJ/mol, respectively. There are no obvious decreases at the third stage of samples combustion process.

1. Introduction

Coal-Based Co-production is the important directions of clean coal technology in future development due to its high efficiency and good environmental performance for electricity generation [1]. A large amount of char are produced in the pyrolysis process, and using this char for combustion and power generation is emerging as viable methods. However, the char is difficult to ignite and burnout as a result of its low volatile and high ash content [2]. At present, there are two different way to solve this problem: preheating combustion [3-4] and co-combustion [5].

Besides, catalytic combustion was regarded as the most promising approaches to raise combustion rate of the pulverized coal and decrease contaminant emission [6]. At present, AAEM (alkaline and alkaline earth compounds) and some transition metal compounds were widely used as catalysts for catalytic combustion of pulverized coal [7]. Gong [8] adopted the differential thermal analysis (DTA) method to study the effects of CeO_2 and Fe_2O_3 on heat release during anthracite combustion. The results of the study showed that ignition temperature decreased and residual carbon decreased with CeO_2 or Fe_2O_3 addition. Zou [9] investigated the influence of MnO_2, CaO, and Fe_2O_3 on the combustion reactivity of anthracite and coal char using thermogravimetric analysis (TGA) and a drop tube furnace (DTF). The DTF results indicated the maximum burnout rate of anthracite and char both increased with catalysts addition. Much research has been aimed to explore the chemical additive effects on coal and char combustion. However, the effects of catalysts on combustion characteristics and kinetics of coal-char blends has not yet been reported.

In this work, effects of Fe_2O_3, CaO, and MnO_2 which were used as catalysts, on coal-char blends combustion were investigated by thermogravimetric analysis (TGA) and differential scanning calorimetry (DSC) technique. The main purpose of the work is to investigate the combustion characteristics and kinetics of coal-char blends with the addition of catalysts.

2. Experimental

The experimental coal used in this work is Shenmu bituminous coal, and the char was produced by a low-temperature pyrolysis method used to process coal. Both the bituminous and char samples were ground to 0.048–0.074 mm. Proximate-ultimate analysis results of both the bituminous and char samples were given in Table 1. The bituminous and char samples were blended in same mass weights (1:1), and named it “CC”.

Table 1. The proximate-ultimate analysis of bituminous and char.

Proximate analysis (%)	Moisture	Ash	Volatile matter	Fixed carbon	
bituminous	12.22	4.51	32.79	60.48	
char	2.18	19.54	3.31	82.35	
Ultimate analysis (%)	Carbon	Hydrogen	Nitrogen	Sulphur	Oxygen
bituminous	75.19	5.87	1.15	0.16	12.61
char	81.47	0.37	0.85	0.59	7.13

The three chemical compounds, Fe_2O_3, CaO, and MnO_2 as catalysts (mass 1.5%), all over 99% in purity with particle size of -0.048 mm, were mixed with coal-char blends by hand for 10 min physically at mortar. Then named them “CC- Fe_2O_3”, “CC- CaO”, and “CC- MnO_2” respectively.

Experiments were performed using TGA/DSC simultaneous thermal analyzer, by which the mass loss curve (TG curve) and heat flow curve (DSC curve) could be obtained. About 10±0.02 mg of samples were placed in an Al_2O_3 crucible and combusted under air atmosphere (50 ml/min), from 30 °C to 850 °C at 15 °C /min.

3. Results and Analysis

3.1. TG-DSC analysis

The characteristic parameters of the samples obtained from TG-DSC curves may be used to compare the effects of catalysts on combustion characteristics of coal-char blends. The TG-DTG and DSC curves of coal-char blends are shown in Figure 1. The ignition temperatures (T_i) and the burnout temperature (T_f) can be determined by TG–DTG tangent method [10]. In addition, an ignition index (C)

and combustion characteristic index (***S***) is also determined using the below mentioned equations to evaluate the ignition and combustion performance of the samples. The higher ***C*** the better the ignition reactivity and the higher ***S*** the better the combustion reactivity.

$$C=\frac{W_{mean}}{T_i^2} \tag{1}$$

$$S=\frac{W_{max}W_{mean}}{T_i^2 T_f} \tag{2}$$

Where; W_{max} is the maximum mass loss rate, W_{mean} is the average mass loss rate from T_i to T_f, T_1 and T_2 are the first and second peak temperature.

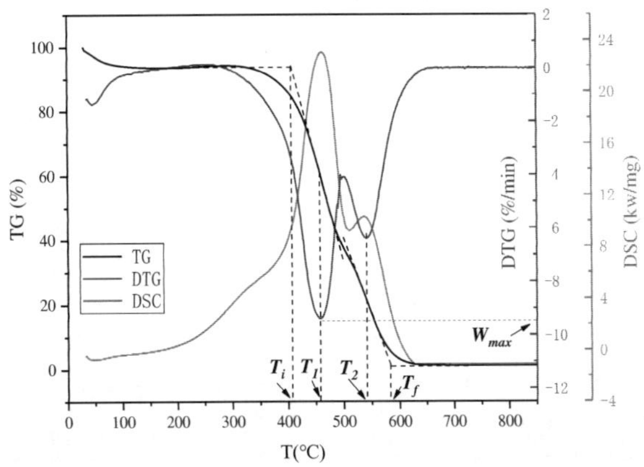

Figure 1. TG-DTG and DSC curves of coal-char blends combustion.

As shown in Figure 1, three mass loss stages were observed. In the first stage, a mass loss less than 8% was observed for combustion temperature up to 200 °C because of the water loss in the bituminous and char. The second stage, ranging from 200 to 500 °C, corresponded to the combustion process of bituminous. And this stage included volatiles combustion and fixed carbon combustion. The third stage, in the temperature range from 500 to 700 °C, corresponded to the combustion process of char. In addition, the weight loss peak temperature was almost equal to the exothermic peak temperature.

A comparison of the reduction degree of the ignition temperature of the four fuels is shown in Figure 2. Besides, the characteristic parameters for samples combustion are listed in Table 2. As shown in Figure 2, with catalysts addition, there are an obvious weight loss peak came before the first weight loss peak in the DTG curve. But the peak of DSC curve with catalysts addition was lower than that the peak of DSC curve without catalysts (CC). This results indicated that Fe_2O_3, CaO, and MnO_2 can improves volatile matter release and reduces kindling point in coal-char blends combustion. Zou [9] considered that the catalytic effect on oxidation reaction occurring on the char particles could reduce the ignition temperature.

As shown in Table. 2, with catalysts addition, the coal-char blends had the lower ignition temperature (T_i) and a lower first peak temperature (T_1). However, the reduction of the second peak temperature (T_1) and the burnout temperature (T_f) was not significant. In addition, the ignition index (***C***) was increased, respectively, by 27% for Fe_2O_3, 6% for CaO, 11.3% for MnO_2, and the combustion characteristic index (***S***) was increased respectively, by 29% for Fe_2O_3, 5% for CaO, 8.3% for MnO_2.

Summarizing, the catalysts had a very meaningful effect on coal-char blends combustion, especially in the second stage. It could improves volatile matter release, and reduces ignition point, promotes char to begin burning under lower temperature. In addition, the catalytic activity of the three catalysts on coal-char blends was (in order): MnO_2<CaO <Fe_2O_3, which is similar to the results of earlier researchers [9].

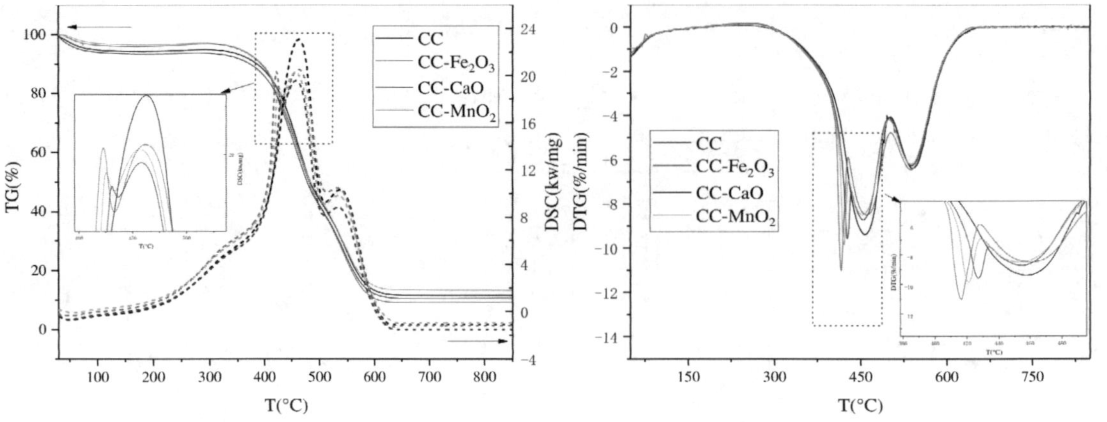

Figure 2. TG-DSC and DTG curves of samples combustion.

Table 2. The characteristic parameters for samples combustion.

Samples	T_i (°C)	T_1 (°C)	T_2 (°C)	T_f (°C)	W_{max} (%·min^{-1})	W_{mean} (%·min^{-1})	C (%·min^{-1}·°C^{-2})	S (%·min^{-1}·°C^{-2})
CC	407.5	457.6	540.6	583.2	9.394	5.992	5.66×10^{-5}	5.81×10^{-7}
CC-Fe_2O_3	391.5	416.6	537.7	581.2	11.031	6.107	7.20×10^{-5}	7.56×10^{-7}
CC-CaO	399.1	427.0	537.8	580.5	9.582	5.874	6.02×10^{-5}	6.09×10^{-7}
CC-MnO_2	395.5	421.1	537.9	582.5	9.837	5.821	6.30×10^{-5}	6.29×10^{-7}

3.2. Kinetic analysis

Due to the reaction mechanisms are quite different in different stages, a sectioning method was adopted to study the kinetics in different stage of combustion processes [11]. In addition, the activation energy (***E***) and Arrhenius constants (***A***) can be calculated using TGA date of single heating rate test by Coats–Redfern method. The coal-char blends combustion was supposed to be governed by the first-order Arrgenius law. And when the heating rate is a constant during combustion, the kinetics of reaction is described as:

$$\frac{d\alpha}{dT}=\frac{A}{\beta}\exp\left(-\frac{E}{RT}\right)f(\alpha) \tag{3}$$

An asymptotic approximation for the resolution of Eq. (3) was used by Coats–Redfern:

$$\ln\left(\frac{g(\alpha)}{T^2}\right)=\ln\left(\frac{AR}{E\beta}\right)-\left(\frac{E}{RT}\right) \tag{4}$$

Where; α is extent of conversion, β is the heating rate, R is the gas constant, $g(\alpha)$ is a function depended by mechanism controlling the reaction.

The general solid-state rate expressions for different mechanisms were given in Table 3.

Table 3. General solid-state rate expressions for different mechanisms

Mechanisms	Symbol	*f(α)*	*g(α)*
Order of reaction			
First order	F_1	1-α	$-\ln(1-\alpha)$
Second order	F_2	$(1-\alpha)^2$	$(1-\alpha)^{-1}-1$
Third order	F_3	$(1-\alpha)^3$	$[(1-\alpha)^{-1}-1]/2$
Diffusion			
One-dimensional diffusion	D_1	$1/2\alpha$	α^2
Two-dimensional diffusion	D_2	$[1-\ln(1-\alpha)]^{-1}$	$(1-\alpha)\ln(1-\alpha)+\alpha$
Three-dimensional diffusion	D_3	$(3/2)(1-\alpha)^{2/3}[1-(1-\alpha)^{2/3}]^{-1}$	$[1-(1-\alpha)^{1/3}]^2$
Limiting surface reaction between both phases			
One dimension	R_1	1	α
Two dimensions	R_2	$2(1-\alpha)^{1/2}$	$1-(1-\alpha)^{1/2}$
Three dimensions	R_3	$3(1-\alpha)^{2/3}$	$1-(1-\alpha)^{1/3}$

According to Table 3, the experimental data of two stage of coal-char blends combustion process were fitted by Eqs.4 by different reaction models. The activation energy (***E***) can be calculated from the slope $\ln(g(\alpha)/T^2)$ versus $1000/T$. Meanwhile, the linear correlation coefficient (R^2) was calculated for different reaction models. The results are shown in Figure 3. The results shows that the line R_2 in Figure 3(a) and F_1 in Figure 3(b) have much higher parallelism, and the correlation coefficients (R^2) of two lines both exceed 0.999. Therefore, the mechanisms of R_2 and F_1 model are suitable for describing the corresponding stage of coal-char blends combustion process.

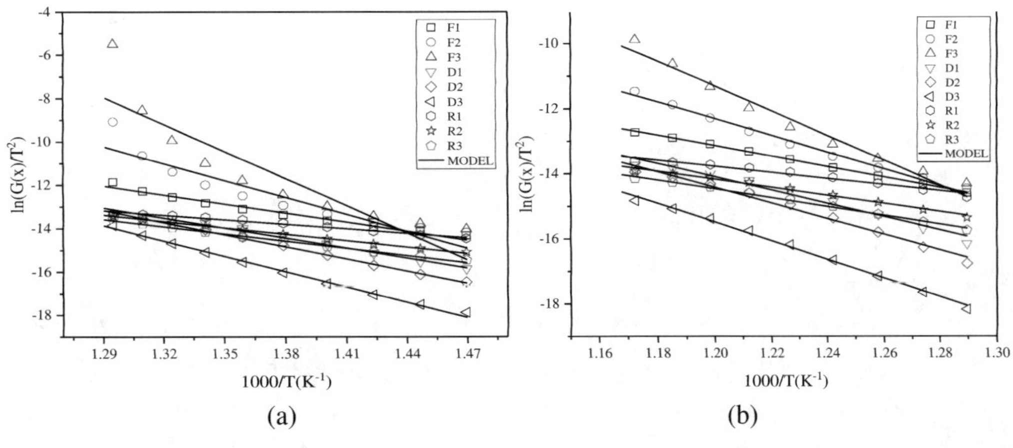

Figure 3. Fitting curves of different reaction mechanisms at the different stage.

The activation energy (**E**) at different stages of samples combustion process are shown in Table 4. With Fe_2O_3 or CaO addition, the activation energy at second stage decreases from 86.0 KJ/mol to 76.92 KJ/mol and 75.12 KJ/mol respectively. However, there are no obvious decreases at the third stage of samples combustion process. It may be concluded that the catalysts exhibits higher catalytic activity at the second stage than the third stage.

Table 4. General solid-state rate expressions for different reaction models.

Samples	Temperature range	*g(α)*	*E* (*KJ/mol*)	R^2
CC	407.5-501.0°C	$1-(1-\alpha)^{1/2}$	86.000	0.999
	501.0-583.2°C	$-\ln(1-\alpha)$	133.897	0.999
CC-Fe_2O_3	391.5-499.8°C	$1-(1-\alpha)^{1/2}$	76.921	0.999
	499.8-581.2°C	$-\ln(1-\alpha)$	128.618	0.999
CC-CaO	399.1-501.8°C	$1-(1-\alpha)^{1/2}$	75.120	0.999
	501.8-580.5°C	$-\ln(1-\alpha)$	127.520	0.999
CC-MnO_2	398.5-498.7°C	$1-(1-\alpha)^{1/2}$	83.329	0.999
	498.7-582.5°C	$-\ln(1-\alpha)$	130.688	0.999

4. Conclusion and Discussion

In this study, the effects of Fe_2O_3, CaO, and MnO_2 on the combustion characteristics and kinetics of coal-char blends were investigated by TGA-DSC.

With catalysts addition, there are other obvious weight loss peak came before the first weight loss peak in the DTG curve, and the coal-char blends had the lower ignition temperature ($\boldsymbol{T_i}$) and a lower first peak temperature ($\boldsymbol{T_1}$). However, the reduction of the second peak temperature ($\boldsymbol{T_1}$) and the burnout temperature ($\boldsymbol{T_f}$) was not significant, and the peak of DSC curve with catalysts addition was lower than that the peak of DSC curve without catalysts (CC). This results indicated that Fe_2O_3, CaO, and MnO_2 can improves volatile matter release and reduces burnout point in coal-char blends combustion.

In addition, the activation energy (***E***) was calculated by Coats–Redfern method. The results indicated that the mechanisms of R_2 and F_1 model are suitable for describing the corresponding stage of coal-char blends combustion process. With Fe_2O_3 or CaO addition, the activation energy at second stage decreases from 86.0 KJ/mol to 76.92 KJ/mol and 75.12 KJ/mol, respectively.

The next work is to investigate the effects of different ratios of catalysts on the combustion and kinetics of coal-char blends. Meanwhile, the release of gases during coal-char blends combustion also will be studied by TGA-DSC-FTIR.

Acknowledgments

This work was sponsored by the The National Key Research and Development Program of China (Contract No. 2017YFB0602002).

References

[1] Li X Y, Xiao Y H and Ren X K 2004 *Clean Coal Technol.* **10** 4-8
[2] Spiro C L, Mckee D W, Kosky P G 1983 *Fuel* **62** 180-184.
[3] Ouyang Z, Zhu J and Lu Q 2013 *Fuel* **113** 122-127
[4] Yao Y, Zhu J and Lu Q 2015 *Energy Fuels* **29** 3985-3991
[5] Sarkar P, Sahu S G, Mukherjee A, Kumar M and Adak A K 2014 *Appl. Therm. Eng.* **63** 616-623

[6] Zou C, Wen L, Zhang S, Bai C and Yin G.2014 *Fuel Process. Technol.* **119** 136-145
[7] Zhang S, Chen Z D, Chen X J and Gong X D 2014 *J. Fuel Chem. Technol.* **42** 166-174
[8] Gong X., Guo Z. and Wang Z 2010 *Energy* **35** 506-511.
[9] Zou C, Zhao J, Li X, and Shi R 2016 *J. Therm. Anal. Calorim.* **126** 1469-1480.
[10] Ma B G, Li X G, Xu L, Wang K and Wang X G 2006 *Thermochim. Acta* **445** 19-22.
[11] Du R, Wu K, Zhang L, Xu D, Chao C, and Zhang B 2017 *J. Therm. Anal. Calorim.* **3** 1-7

Integration Research on Gas Turbine and Tunnel Kiln Combined System

Shi Hefei, Ma Liangdong, Liu Mingsheng
Dalian University of Technology, China

Abstract. Through the integrated modeling of gas turbine and tunnel kiln combined system, a thermodynamic calculation method of combined system is put forward, and the combined system operation parameters are obtained. By this method, the optimization of the combined system is analyzed and the optimal configuration of the gas turbine is calculated. At the same time, the thermal efficiency of the combined system is analyzed, and the heat distribution and thermal efficiency of the system before and after the improvement are explained. Taking the 1500 kg/h ceramic production as an example, pointed out that if the tunnel kiln has a gas turbine with a power of 342 kw. The amount of electricity of the combined system that produced per unit volume of the fuel which consumes more than it used to will be 7.19 kwh, the system thermal efficiency will reach 57.49%, which higher than the individual gas turbine's cycle thermal efficiency 20% at least.

1. Introduction

In the actual sintering process, ceramic products have strict requirements for temperature, and the burning temperature of the products is generally about 1200 °C [1]. In order to reduce the combustion temperature of natural gas, a large amount of excess air is needed. Therefore, the exhaust gas from the flue pipe contains a large amount of air have not involved in combustion reactions, the exhaust gas temperature is about 250 ℃[1], this part of the gas is discharged into the atmosphere, which take away a lot of heat, lead to the reduction of kiln heat utilization rate and the waste of the fuel energy. In order to reduce the amount of oxygen content that has not been involved in the combustion process, a gas turbine model in a tunnel kiln has been built. In this technique, first burn natural gas and excess air[2], the combustion produces 1000 ℃ and over high temperature exhaust gas into the turbine, which for gas turbine electric power generation, and then put the medium temperature oxygen enriched steam that exhausted from the turbine into the kiln, burn with the supplementary natural gas and heat the ceramic products. Finally, the amount of oxygen in the exhaust gas that exhausted from the kiln is greatly reduced, the heat loss of cxhaust gas is reduced, and the energy efficiency of the kiln is improved. Gas turbine and tunnel kiln combined system transform natural gas energy into electrical energy firstly, and then for heat supply, thermal efficiency of combined cycle system can reach more than 55%.

2. Thermodynamic model of gas turbine tunnel kiln combined system

Figure 1 is the principle diagram of gas turbine tunnel kiln combined operation: a part of the fuel and excess compressed air mix and burn in the combustion chamber[3], high temperature exhaust gas produced (1200 ℃~1400 ℃) ,driven the gas turbine to generate electricity firstly, then the medium temperature oxygen enriched gas (about 500 ℃~600 ℃) was brought into the tunnel kiln, heat the material to the specified target process temperature together with the supplementary fuel.

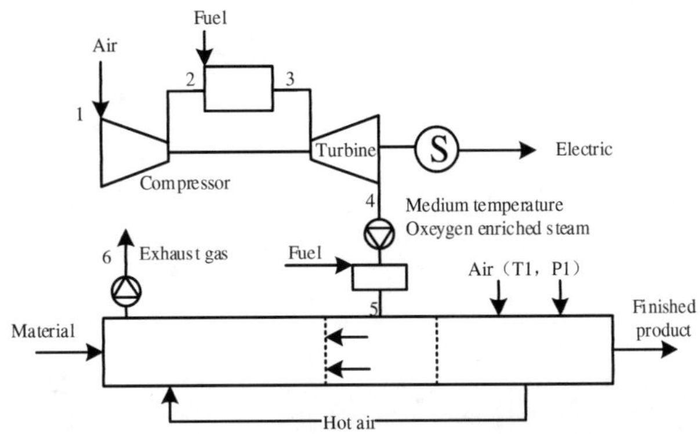

Figure 1. Schematic diagram of combined operation of kiln and gas power generation.

The thermodynamic process of a gas turbine tunnel kiln combined system is shown in Figure 2: the operation of the combined system is 1-2-3-4-5-6-1. The compressor draws air from the atmosphere and compresses it. Then the high pressure air is sent to the combustion chamber, mixed with fuel and burned at constant pressure to produce high temperature gas. The combustion temperature reaches T3. the gas enters the gas turbine to expand, and then drives the impeller and out puts shaft power to exhaust the gas and the temperature of the exhaust the gas is T_4. The exhausted steam is fed into the kiln and mixed with fuel and burned, this process is a constant pressure process 4-5, heat transfer from burning gas to material, thermodynamic process is 5-6, it is a constant pressure process. After the heat transfer, the gas temperature drops to T_6. Exhaust gas is discharged into the atmosphere from the flue. The process (6-1) is exothermic under constant pressure.

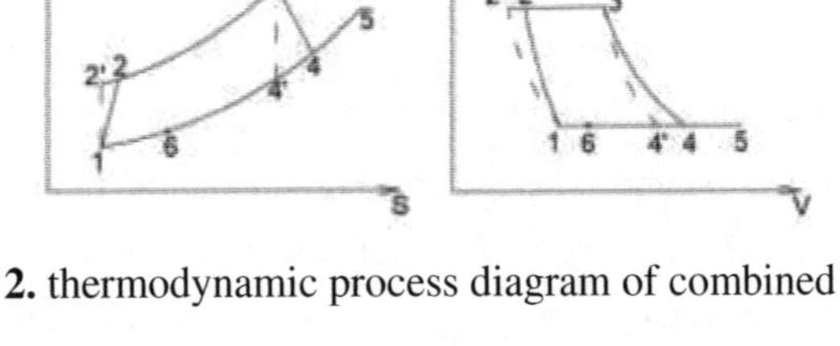

Figure 2. thermodynamic process diagram of combined system.

IOP Conf. Series: Earth and Environmental Science **133** (2018) 012024 doi:10.1088/1755-1315/133/1/012024

According to the formula of heat efficiency, the cycle thermal efficiency of gas turbine tunnel kiln combined system can be expressed as:

$$\eta=\frac{w_0}{q_1}=\frac{w_{3\text{-}4}-w_{1\text{-}2}}{q_1}=\frac{(h_3\text{-}h_4)\text{-}(h_2\text{-}h_1)}{(h_3\text{-}h_2)+(h_5\text{-}h_4)-(h_5\text{-}h_6)} \tag{1}$$

The specific heat capacity is taken as a fixed value, and the upper form can be rewritten as:

$$\eta=\frac{c_{p0}(T_3\text{-}T_4)\text{-}c_{p0}(T_2\text{-}T_1)}{c_{p0}(T_3\text{-}T_2)+c_{p0}(T_5\text{-}T_4)\text{-}c_{p0}(T_5\text{-}T_6)}=\frac{(T_3\text{-}T_4)\text{-}(T_2\text{-}T_1)}{(T_3\text{-}T_2)\text{-}(T_4\text{-}T_6)} \tag{2}$$

The cycle heat efficiency of an individual gas turbine is: $\eta'=\frac{(T_3\text{-}T_4)\text{-}(T_2\text{-}T_1)}{(T_3\text{-}T_2)}$ [2]

3. Heat balance calculation of combined system

The heat balance calculation logic of the combined system is shown in Figure 3:

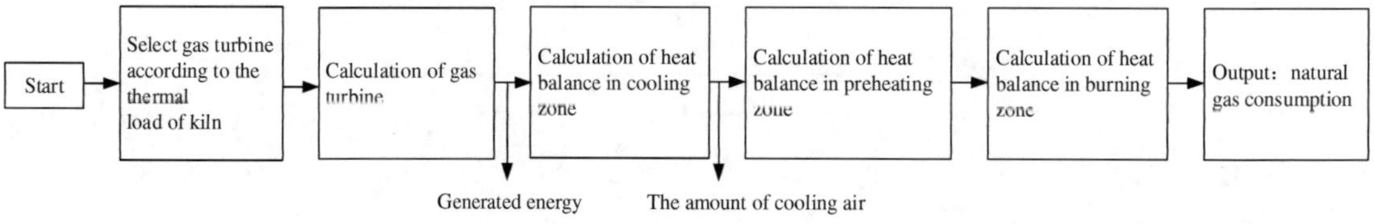

Figure 3. logic diagram of thermodynamic calculation of gas turbine tunnel kiln combined system

3.1. Heat balance calculation of cooling zone

The heat income of the cooling zone is equal to the heat output, and the heat balance of the cooling zone is shown in figure 4.

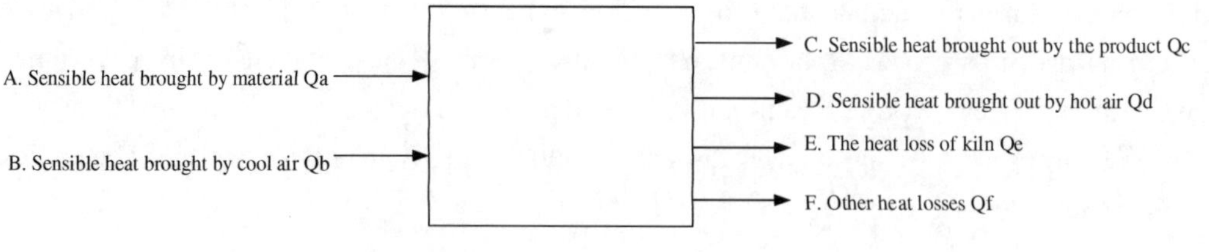

Figure 4. Schematic diagram of heat balance of cooling zone

The heat balance formula of cooling zone is [2]:

$$Q_a+Q_b=Q_c+Q_d+Q_e+Q_f \tag{3}$$

According to the formula of heat balance, the cold air volume can be obtained by solving the equation. In the same way, a balance equation between the preheating zone and the firing zone can be obtained according to the equilibrium equation of heat budget. Thus the consumption of natural gas is calculated.

4. Combined system energy evaluation index

4.1. Cyclic thermal efficiency of combined system

The thermodynamic cycle of the combined system shows that the cycle thermal efficiency of the gas turbine tunnel kiln system is obviously higher than that of the single gas turbine when the gas turbine tunnel kiln combined system is be use[4]d, from the above calculation we can see that

$$\eta=\frac{(T_3-T_4)-(T_2-T_1)}{(T_3-T_2)-(T_4-T_6)} \tag{4}$$

The higher the cycle thermal efficiency of the combined system, the less energy wasted and the more energy saved, so cycle thermal efficiency of the system is an important indicator of the evaluation of the integrated system.

4.2. The amount of electricity that produced per unit volume of the fuel which consumes more than it used to of the combined system

The generating capacity of the individual gas turbine is 3 kwh/m³~4 kwh/m³[5],After the integration of gas turbine and tunnel kiln, the natural gas consumption has increased by ΔV ,The electrical energy produced is W_e. Therefore, from the point of view of the combined system. The amount of electricity that produced per unit volume of the fuel which consumes more than it used to can be calculated by the next formula:

$$e=\frac{W_e}{\Delta V} \tag{5}$$

The higher the amount of electricity that produced per unit volume of the fuel which consumes more than it used to, the higher efficiency of the high-grade energy will be produced by the combined system.so The amount of electricity that produced per unit volume of the fuel which consumes more than it used to be an important indicator of the evaluation of the integrated system.

5. Analysis of calculation results

Taking the weight of 1500 kg into the kiln per unit time as an example, the thermodynamic performance can be obtained before and after the integration:

Table 1. Comparison of the integrated results with individual kiln

items	before integrating	after integrating
Material inlet quality/kg	1500	1500
Material moisture Content	2%	2%
High temperature coefficient of kiln	0.85	0.85
The temperature of natural gas entering the kiln/℃	20	20
Compressor outlet pressure/MPa	—	20
Excess air coefficient	1.63	2.43

The temperature of the material entering the kiln/℃	20	20
The temperature of material entering the firing zone/℃	1043	1049
Sintering temperature of material /℃	1200	1200
Material output temperature/℃	80	80
Exhaust gas temperature /℃	250	250
Hot air temperature/℃	200	200
Fuel combustion temperature/℃	1413	1413
Total consumption of natural gas/m^3	144.9	192.57
The consumption of natural gas in gas turbine/m^3	—	99.34
The consumption of natural gas in kiln/m^3	144.9	93.23
Low heating value of the natural gas /kJ/m^3	36,000	36,000
A total amount of consumed heat/kJ	5,220,862.92	6,938,445.45
The heat consumption of gas turbine /kJ	—	3,579,299.67
The heat consumption of kiln/kJ	5,220,862.92	3,359,145.78
Fume Exhaust Volume /m^3	9.929.398	10,026.31
Heat loss due to exhaust gas	67.18%	53.31%
The amount of electricity that produced per unit volume of the fuel which consumes more than it used to /kwh/m^3	—	7.19
Cycle thermal efficiency	—	57.49%
The price of natural gas/RMB/m^3	2.8	2.8
The total price of natural gas/RMB	405.72	539.20
generated energy /kwh	—	342.65
electricity price /RMB/kwh	0.8	0.8
Electricity generation income/RMB	—	274.12

Table 2. Gas turbine operating parameters (pressure ratio is 20)

Compressor inlet temperature/ C°	Compressor outlet temperature/ C°	Compressor efficiency	Combustion chamber efficiency	Turbine efficiency	Turbine inlet initial temperature/C°	Exhausted temperature/ C°
20	300	0.85	0.99	0.88	1226	635

Compared with the amount of the natural gas consumption of kiln, the gas turbine kiln tunnel combined system increased 47.67 m^3. But it also increased 342.65 kwh of electricity, equivalent to 1 m^3 of natural gas made 7.19 kwh of electricity. When the gas turbine operates alone, 1m^3 natural gas only generate 3.45 kwh of power. After integration, the efficiency of gas generation is equivalent to 2.08 times of that before integration. Based on the load of the kiln, thermal efficiency of combined cycle system reached 57.49%, 23.03% higher than the cycle thermal efficiency of individual gas turbine, efficiency of the combined system is improved greatly, the energy utilization improved, energy saving effect is remarkable.

6. Conclusion

An energy saving technology by putting gas turbine in front of tunnel kiln is put forward. A calculation model of heat balance of gas turbine tunnel kiln is built. By this method, the operating parameters of the combined system that can satisfy the firing temperature of the kiln are calculated, including gas turbine operating parameters and kiln operation parameters.

Acknowledgments

This work is supported by National Key Research and Development Project of China No. 2017YFC0704100 (entitled New generation intelligent building platform techniques).

7. Reference

[1] Liu Zhenqun, Yang Shuzi and Huang Bingjun 1978 *Ceramic industrial kiln* (Beijing: China Architecture& Building Press)

[2] Shen Weidao, Jiang Zhimin and Tong Jungeng 1982 *Engineering Thermodynamics* (Beijing: Higher Education Press)

[3] S Kaya, K Küçükada, E Mançuhan.Model-based optimization of heat recovery in the cooling zone of a tunnel kiln[J]. Applied Thermal Engineering, 2008, 28(5):633-641.

[4] Katsuki M., Hasegaw T .The science and technology of combustion in ighly preheated air[A].27th Symposium (Internation)on combustion[C].Pittsburgh:the Combustion Institute 1998 , 3135 -3146 .

[5] Ana Mezquita, Juan Boix, Eliseo Monfort, Gustavo Mallol. Energy saving in ceramic tile kilns: Cooling gas heat recovery.[J].Applied Thermal Engineering, 65(2014):102-110.

[6] E.A. Abdelaziz, R. Saidur, S. Mekhilef, A review on energy saving strategies in industrial sector. Renewable & Sustainable Energy Reviews, 15 (2011) 150-168.

[7] MDG Carvalho, M Nogueira. Improvement of energy efficiency in glass-melting furnaces, cement kilns and baking ovens[J]. Applied Thermal Engineering, 1997, 17(8–10):921-933.

[8] EN 14411. Ceramic tiles. Definitions, classification, characteristics and marking; 2003.

[9] M. Kaden, Controlling the residual oxygen content of kiln gases, ZI Int. 58 (8)(2005) 26-32.

IOP Conf. Series: Earth and Environmental Science **133** (2018) 012025 doi:10.1088/1755-1315/133/1/012025

A study on post impingement effects of urea-water solution spray on the heated wall of automotive SCR systems

G M H Shahariar, M K A Wardana and O T Lim*

School of Mechanical Engineering, University of Ulsan, Ulsan, 680-749, Korea

E-mail: otlim@ulsan.ac.kr

Abstract. The post impingement effects of urea-water solution spray on the heated wall of automotive SCR systems was numerically investigated in a constant volume chamber using STAR CCM+ CFD code. The turbulence flow was modelled by realizable k-ε two-layer model together with standard wall function and all y+ treatment was applied along with two-layer approach. The Eulerian-Lagrangian approach was used for the modelling of multi phase flow. Urea water solution (UWS) was injected onto the heated wall for the wall temperature of 338, 413, 473, 503 & 573 K. Spray development after impinging on the heated wall was visualized and measured. Droplet size distribution and droplet evaporation rates were also measured, which are vital parameters for the system performance but still not well researched. Specially developed user defined functions (UDF) are implemented to simulate the desired conditions and parameters. The investigation reveals that wall temperature has a great impact on spray development after impingement, droplet size distribution and evaporation. Increasing the wall temperature leads to longer spray front projection length, smaller droplet size and faster droplet evaporation which are preconditions for urea crystallization reduction. The numerical model and parameters are validated comparing with experimental data.

1. Introduction

In order to meet the progressive compact emission legislation, such as Euro 6, engine manufacturers are being forced to install exhaust after treatment devices to meet the raising challenge. The selective catalytic reduction (SCR) is the most felicitated automotive exhaust gas after treatment solution for nitrogen oxides (NOx) emissions reduction which is capable of meeting most of the emission standards [1]. Ammonia (NH_3) gas reacts with the flowing exhaust gas into the SCR device and converts the NOx element into nitrogen (N_2) gas and water (H_2O) [2]. Though NH_3 is a noxious chemical component and harmful for human heath, urea is used as the precursor of NH_3 gas, which can be easily handled and transported [3]. Urea-water solution (32.5% urea by weight) known as Ad-Blue is injected into the exhaust gas stream by means of an injector, and then urea is decomposed to ammonia (NH_3) gas [4]. Urea decomposition is very slow process and catalyst is required, as a result isocyanic acid reacts with undecomposed urea and produces complex polymers such as melamine, biuret, ammeline, triuret, cyanuric acid and ammelide [5]. According to the study of Xu et. al [6] solid deposits are formed below 300° C temperature in SCR system. These by-products (urea deposits) are accumulated on the injector tip, exhaust pipe wall, catalyst surface, mixing element and reduce the catalyst activity, increase engine backpressure, thus lower the NOx conversion rate and gradually make the system ineffective. When the exhaust temperature is more than 350° C, the deposits are vaporized gradually, but thermal decomposition of deposits are very slow, the rate is about 2 mg/min at 873 K temperature [7]. So optimization of the UWS spray is the most feasible way to minimize deposit formation.

For acquiring this specific purpose Computational fluid dynamics (CFD) analysis can be the effective tool for the designers. Application of automatic meshing approach can be operative here to assure the maximum possible prediction accuracy within a short time frame. In the purpose of urea-SCR simulation various modelling approaches such as ammonia homogenization [8], deposit formation are established. Spray-wall interaction is the most complicated and significant phenomena, that is modelled by Bai-Gosman [9] and others. For the proper understanding of UWS droplet impingement process and mechanisms underlying behind this, an intensive investigation is required in a constant volume environment, where only spray wall impingement take place. This paper presents a three dimensional numerical modelling and analysis of UWS spray wall impingement to investigate the post impingement effects on the heated wall of SCR system using STAR CCM+. The numerical simulation measured the axial and radial distance of UWS spray after impinging on the heated wall. The numerical visualization and measurements are compared and validated with the experimental results. Moreover, the simulation estimated two crucial parameters; UWS droplet size distribution and UWS droplet evaporation rate after the start of injection using the specially developed user defined functions (UDF). A detailed analysis was carried out to gain a deep understanding about the post impingement effects and the causes beyond deposit formation in SCR systems.

2. Simulation conditions

The urea-water solution (UWS) known as Ad-blue was sprayed on the heated wall by means of a single hole pressure swirl injector in a non-reacting environment.

Table 1. Spray simulation case setup and conditions.

Parameter	Value
Injected liquid	UWS (32.5% wt. urea)
Injection duration	10 ms
Height from the wall, h	30 mm
Injection pressure, P_{inj}	0.47 Mpa
Injector nozzle hole diameter	0.55 mm
Injection angle, θ	90 ^{0}
Half cone angle, Φ	29 ^{0}
Wall temperature, Tw	338, 413, 473, 503, 573 K

3. Numerical modelling

The interaction between each particle or parcel and with the continuous phase are modelled in Eulerian-Lagrangian approach using the incompressible and unsteady Reynolds averaged

NavierStokes (RANS) equations for energy, mass, momentum and species. The realizable k-ε model is applied for the modelling of the turbulent flow because it has improved ability over standard k- ε model in case of turbulence quantities estimation. The primary spray atomization for the thin liquid sheet produced at the nozzle tip, created by the pressure swirl atomizer is modelled by Linearized Instability Sheet Atomization (LISA) model [10]. For modelling the droplet size distribution an empirical function known as Rosin-Rammler distribution is used [11]. The collision between the droplets are estimated by the NTC (No Time Counter) collision detection algorithm [12]. The atomized droplet trajectories are estimated by Newton's second low of motion that states the momentum rate change is equal to the summation of forces acting on the particle. The quasi steady evaporation rate of each component is described as

$$m_{pi} = -\zeta g * A_s \ln(1+B) \tag{1}$$

Here, B denotes Spalding transfer number, as denotes the surface area of droplet, g* denotes the mass transfer conductance, ζi represents the fractional mass transfer rate. The commercial simulation software STAR CCM+ is used for solving the governing equations. The second order upwind scheme is executed for multiphase simulations, because it has better accuracy than the first order scheme. In droplet wall interaction modelling Leidenfrost temperature effect is implemented using a specially developed user defined function (UDF) subroutine. Two more UDF are implemented to solve the droplet size distribution and droplet evaporation. All y+ treatment was applied along with two layer approach. The residuals for energy, turbulent kinetic energy, turbulent dissipation rate, momentum, temperature, continuity and each species were inspected until stabilization to trace the convergence. In this work, automatic meshing technique was used with a base size of 5 mm volume mesh type. Two prism layers with 2 mm thickness were used for resolving the near wall region flow accurately. Sensitivity analysis was done for mesh base size 4 mm and 3 prism layers and the other settings were remained same.

4. Results and discussion

During the simulation study three crucial parameters (spray development after impingement, droplet SMD and droplet evaporation) have been investigated and analysed to reveal the post wall impingement effects in diesel SCR system.

4.1. UWS spray wall impingement at low wall temperature (338 K)

UWS spray development after impinging on the heated wall for a wall temperature 338 K is seen in figure 1. Figure 2 illustrates the comparison of the numerical and experimental UWS spray visualization of UWS spray wall impingement for a wall temperature 338 K and for three separate time intervals; t = 5,8 &10 ms after the start of injection. The droplets of the numerical images are coloured based on their velocities. The acting wall temperature is lower than the boiling point temperature of water. After impingement UWS spray droplets are adhered on the wall surface and urea deposits are formed, if the temperature of the wall is lower than 450 K. The hollow cone spray produced by the current numerical simulation is analogous with the experimental visualization. The UWS spray projected radially after impacting on the heated wall, which is clearly seen in current simulation and experimental visualization. A detailed look onto the measurements of axial and radial distance of UWS spray after impinging on the heated wall is presented in figure 3. Axial distance travelled by the UWS spray after impinging on the heated wall is in good agreement at all the three analysis; radial projection distance measurements are harmonious with the experiment but Abu-Ramadan simulation measurements are deviated.

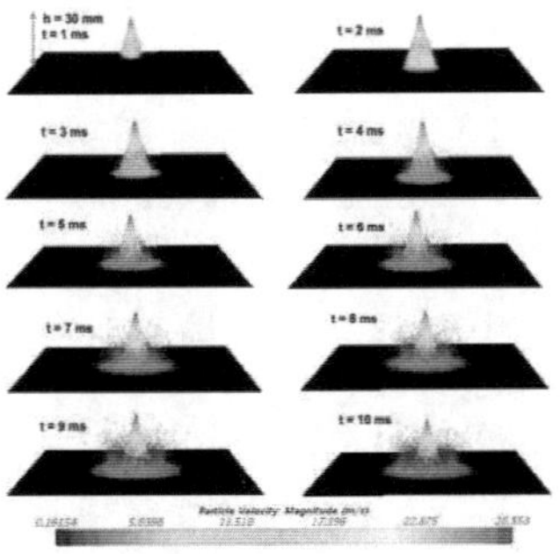

Figure 1. UWS spray development after impinging on the heated wall, at Tw = 338 K.

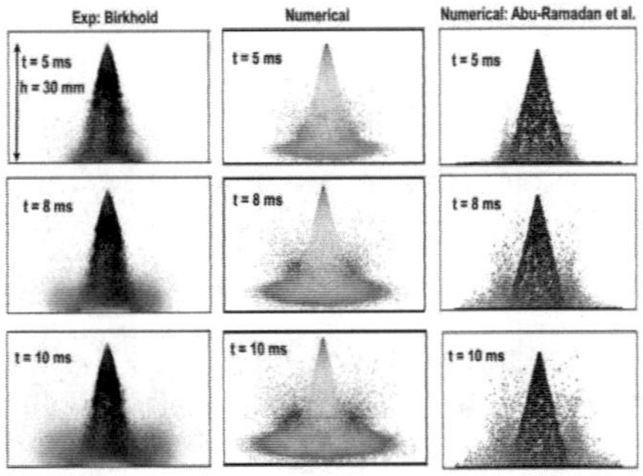

Figure 2. Comparison between the visualization of current simulation, Birkhold experiment and Abu-Ramadan et al. simulation, at Tw = 338 K.

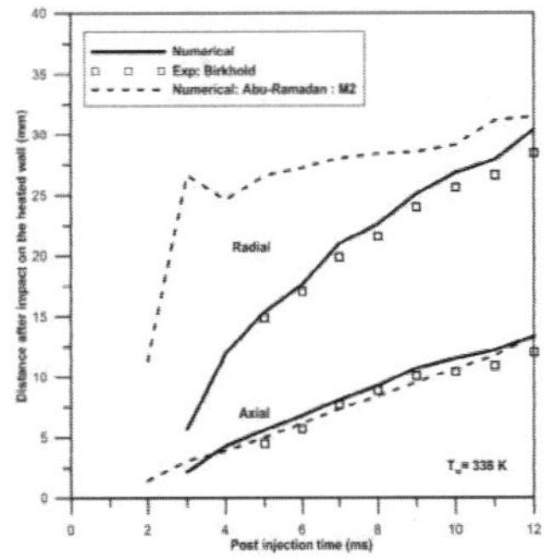

Figure 3. Comparison between the measurement current simulation, Birkhold experiment, and Abu-Ramadan et al. simulation, at Tw = 338 K.

4.2. UWS spray wall impingement at high wall temperature (573 K)

Figure 4 shows the spray impingement phenomena after impinging on the heated wall for the wall temperature 573 K. Figure 5 shows the comparison of the numerical and experimental UWS spray visualization for three separate time frames; t = 5,8 &10 ms after the start of injection. A little deviation stands between the experimental and current numerical visualization. The splashed/rebounded droplets in the current simulation are analogous with the experiment. In the experimental images vaporized wavy region can be seen, but Lagrangian particle tracking unable to display those phenomena. In the current simulation study, the spray development area seems smaller than the experiment, this is because the evaporated droplets are not shown in simulation visualization. Figure 6 shows the quantitative comparison of the axial and radial distances of UWS spray after impinging on the heated wall. The current simulation underestimates the axial and radial distances in comparison with the experiment, after impinging on the heated wall. This is because the simulation visualizations are not showing the evaporated droplets, hence the axial and radial distances reduced, but the current simulation results are identical with the experimental measurements.

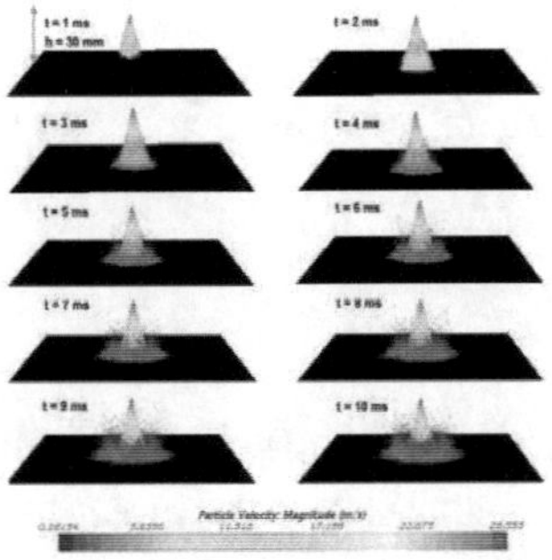

Figure 4. UWS spray development after impinging on the heated wall, at Tw = 573 K.

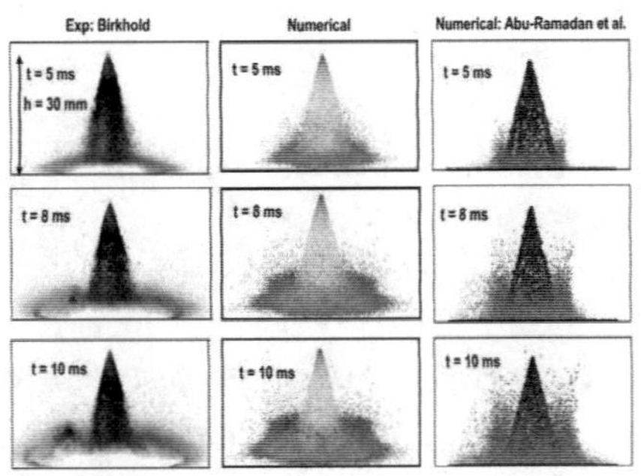

Figure 5. Comparison between the visualization of current simulation, Birkhold experiment and Abu-Ramadan et al. simulation, at Tw = 573 K.

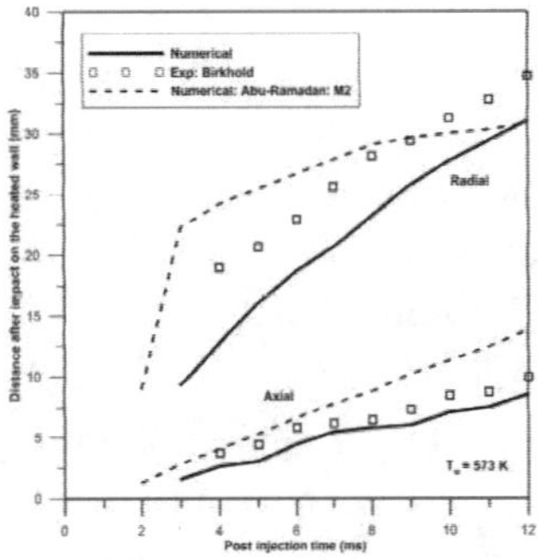

Figure 6. Comparison between the measurement of current simulation, Birkhold experiment and Abu-Ramadan et al. simulation, at Tw = 573 K.

4.3. Droplet size distribution for different wall temperature

Figure 7 presents the sauter mean diameter (SMD) of the UWS droplets after the start of injection, for wall temperature 338, 413, 473, 503 & 573 K. During the UWS spray impinging on the heated wall, small size droplets evaporate faster that leads to form higher active elements during the same time, therefore the exhaust gas and SCR reductant mixes uniformly, thus deposit formation is reduced. Simulation studies reveal that the droplet SMD is almost same for the wall temperature 338, 413, 473 and 503 K, so urea solid deposit formation risk is still high until the wall temperature 503 K. The SMD is stabilized quickly for high wall temperature (573 K). The maximum value is 98 μm for 338 K wall temperature and 51 μm for 573 K wall temperature, which is about half of the first case. Shorter liquid phase exhibit period is better for faster droplet evaporation; the value is 1.3 ms for 338 K wall temperature and 0.3 ms for 573 K wall temperature.

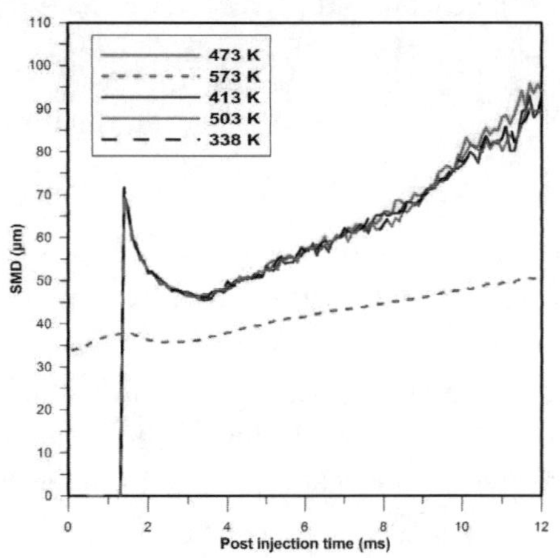

Figure 7. Droplet size distribution of UWS at, Tw = 338 K, 413 K, 473 K, 503 K & 573 K.

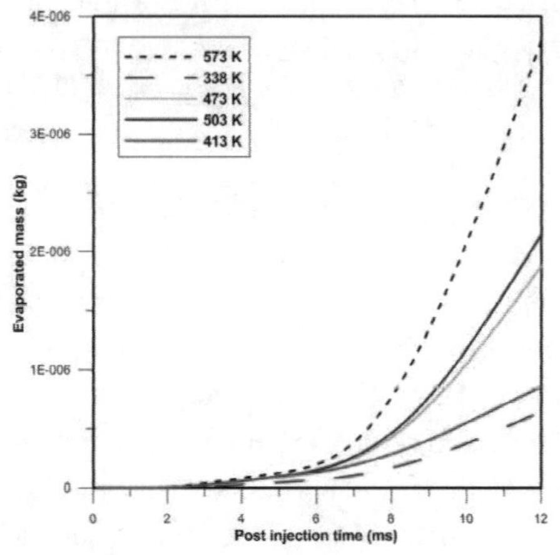

Figure 8. Evaporation rate of UWS at, Tw = 338 K, 413 K, 473 K, 503 K & 573 K.

4.4. Evaporation behavior of UWS

Rapid evaporation of UWS is recommended for the reduction of urea deposits formation. Figure 8 shows the evaporation rates of UWS for wall temperature 338, 413, 473, 503 & 573 K. It is seen that, the UWS droplets evaporation rate increases gradually for lower wall temperature to higher wall temperature and it goes to the peak for 573 K wall temperature. The highest value is 1.65e-6 kg for 573 K wall temperature and 4.02e-7 kg for 338 K wall temperature. Therefore, high temperature operation is advantageous for the reduction of deposit formation.

5. Conclusion

In order to address concerns about the effects of wall impingement of UWS on the heated wall of SCR systems, a numerical study has been conducted using STAR CCM+ CFD code. The spray wall impingement, droplet size distribution and droplet evaporation rates are analysed and the effects of wall impingement on urea deposits formation are investigated under different ambient conditions. The simulation results are compared with Birkhold experiment [13] and Abu-Ramadan et al. simulation [14]. Following conclusions are drawn from the study.

(1) The wall temperature significantly influences the development of UWS spray after impinging on the heated wall. The development of spray front (radial projection) is larger at high wall temperature

and axial length is smaller i.e larger spray distribution area, which is better for efficient mixing and evaporation thus lower urea deposits formation and vice versa for low wall temperature.
(2) The predicted SMD is rather small in high wall temperature and small droplets evaporated completely.
(3) The evaporation rate is very slow at low wall temperature in comparison to high wall temperature, this is because the simulation is predicted much higher size droplets and lower radial projection distance at low wall temperature operation.

Acknowledgments
This research was financially supported by the Global-Top Project of Korea Ministry of Environment (2016001420002, Development of the PM·NOx purifying system and the core technology).

6. References

[1] Johnson T. 2016 SAE Int. J. Engines 9(2):1258-1275
[2] Selective catalytic reduction. (Online), 2009 (cited 2017-02-10), Available from: http://www.dieselnet.com/tech/cat_scr.php
[3] Salib R. and Keeth R. 2003 Proce. of Mega Symposium
[4] Fritz, A. and Pitchon V. 1997 Appl.Catalysis B: Environ. 13(1), 1-25
[5] Birkhold F. Meingast U. Wassermann P. and Deutsch-mann, O. 2006 SAE Tech. Paper 2006-*01-0643*
[6] Xu L. Watkins W. Snow R. Graham G. et al. 2007 SAE Tech. Paper 2007-01-1582
[7] Way P. Viswanathan K. Preethi P. Gilb A. et al. 2009 SAE Tech. Paper 2009-01-0633
[8] Fischer S. Bitto R. Lauer T. Krenn C. et al. 2012 SAE Int. J. Engines 5(3):1443-1458
[9] Bai C. and Gosman A. 1995 SAE Tech. Paper 950283
[10] Senecal P.K. Richards K.J. Pomraning E. Yang T. Dai M.Z. McDavid R.M. Patterson *M.A.,Hou, S. and Shethaji T. 2007. SAE Tech. Paper 2007-01-0159*
[11] STAR CCM+ version 11.04 Theory Guide, 2016
[12] Schmidt D.P. and Rutland C.J. 2000. J Comp. Phy.
[13] Birkhold F. 2007 Untersuchung der Einspritzung von Harnstoffwasserlösung. Shaker
[14] Abu-Ramadan E. Saha K. and Li X. 2012 SAE Tech. Paper 2012-01-1301

Study on the ecosystem construction of using ecopath model in inland waterway

Junjie Zhao, Jing Bai*, Lu Zhang, Ning Wang and Youping Shou
Key Laboratory of Environmental Protection Technology on Water Transport, Ministry of Transport, Tianjin Research Institute of Water Transport Engineering, Tianjin , 300456, China

Abstract. In this paper, Ecopath with Ecosim 5.1 software is used to simulate the constructed water ecosystem of inland waterway. According to the characteristics of feeding relationship, the ecopath model of water ecosystem is divided into seven functional groups: phytoplankton, hydrophyte, zooplankton, herbivorous, omnivorous, polychaetes and detritus. By analyzing the important ecological parameters of the ecosystem, such as biomass, biomass / biomass, consumption / biomass, trophic level and ecological nutrient conversion efficiency, the software integrates the energy flow process of the ecosystem, the ratio of the total net primary production and the sum of all respiratory flows is 1.314, it's indicating that the ecosystem is equilibrium. The research method of this paper can be widely used to evaluate the stability of the ecosystem of the domestic river.

1. Introduction

Excavation engineering of inland waterway will damage the fish, benthos and phytoplankton in the water, and then affecting the entire water ecosystem. It's a hot topic of how to repair and rebuild the water ecosystem, the research of inland waterway is located in Punan Village ,Dapu Town,Yixing City,Jiangsu Province (119°49'E， 31°22'N), the area faces Taihu lake on the east, which belongs to the network of rivers, flat, thick rivers, bidirectional flow. The construction of the ecosystem of inland waterway is based on the common software of Ecopath with Ecosim5.1 developed by ICLARM (International Center for Living Aquatic Resources Management). The software uses the nutrition dynamics principle to construct the water ecosystem structure directly, describe the energy flow and determine the energy balance model of the ecological parameters, and estimate the biomass and food consumption of the different groups in the aquatic ecosystem[1-3].

2. Software theory and method

The ecosystem defining by Ecopath model is maded up of a series of ecologically relevant functional components (group or box) that can include organic debris, plankton, a fish, a particular age group or a type of ecosystem characterization (such as feeding) in same species, all functional components can basically cover the whole process of energy flow in the ecosystem [4]. According to the thermodynamics principle, the Ecopath pattern defines the energy input and output balance of each biological function group in the system, the energy balance is expressed as: production - predation death - other natural death - output = 0.

The prerequisite for using the software is to assume that the biomass of each biological group is stable in a year, so the production rate for this year = predation + non-predation loss + output. The consumption rate is related to the consumption rate of the predator and the diet composition, and the natural death and output can be calculated respectively from the ecotrophic efficiency and the fish catch and sedimentation rates. According to the characteristics of the relationship with feeding habits, the Ecopath model of the aquatic ecosystem is divided into seven boxes, Phytoplankton, Hydrophyte, Zooplankton, Herbivorous Fishes, Omivorous Fishes, Polychaetes and Detritus. The division of boxes can basically cover the main flowing process of biological energy in aquatic ecosystems in the region.

3. Determination of parameters

In the computer model, the biology of each unit is required, and at least three items should be entered in the following five data,

Biomass (B), production of unit biomass (P/B), metabolic consumption per unit biomass (C/B), ecological efficiency (EE, the parameters of any of the various energy flow parameters between the trophic level and the trophic level), gross efficiency(GE). P/B and Q/B are based on the actual situation, the relevant domestic data estimates and checking through the Fishbase network (www.fishbase.org)[5,6].Ecological efficiency is a more difficult parameter to obtain, in the input parameters of the Ecopath model, the EE of most functional groups is usually unknown. One of the debugging processes of the Ecopath model is to adjust the value of all EE to less than 1, so that the flow of energy in the whole system to maintain balance, so as to obtain a reasonable ecological parameters of the ecological system (EE greater than 1 is wrong, indicating that the energy flow equilibrium condition in the system is destroyed, because the using of any organism in the system can not be greater than the energy output of the organism itself [7].

The basic input of the model is shown in figure 1.

3.1. Determination of biomass

The biomass of the three functional groups of the omnivorous fishes, herbivorous fishes and the hydrophyte are $40g/m^2$, $50g/m^2$ and $500g/m^2$, the local data of the zooplankton, polychaetes, phytoplankton and detritus are $1.4g/m^2$, $57.22g/m^2$,$7g/m^2$ and $10g/m^2$.

3.2. Determination of production of unit biomass (P/B)

The P/B values for fishes are equal the total mortality Z.

$$Z = \ln X \tag{1}$$

X is the survival rate of fish, omnivorous fish is 0.75, herbivorous fish is 0.5), the P/B values for omnivorous fish and herbivorous fish are 0.3 and 0.7, the P/B values for zooplankton, polychaetes, phytoplankton and hydrophyte are 20, 1.1, 50 and 1.2 [8-10].

3.3. Determination of metabolic consumption per unit biomass (C/B)

IOP Conf. Series: Earth and Environmental Science **133** (2018) 012026 doi:10.1088/1755-1315/133/1/012026

C/B values for using (P/B)/(C/B)= gross efficiency ,the general gross efficiency is between 0.1-0.3, in which phytoplankton and hydrophyte belong to autotrophic organisms, the value of C/B is 0.

Input data

	Group name	Habitat area (fracti	Biomass in hab. area (Wet weight (t/km?)	Production / biomass (/year)	Consumption / biomass (/year)	Ecotrophic efficiency	Production / consumption
1	Omivorous	1.000	40.000	0.300	1.200		
2	Herbivorous	1.000	50.000	0.700	2.800		
3	Zooplankton	1.000	1.400	20.000	62.000		
4	Polychaetes	1.000	57.220	1.100	14.000		
5	Phytoplankton	1.000	7.000	50.000			
6	Hydrophyte	1.000	500.000	1.200			
7	Detritus	1.000	10.000				

Basic input | Diet composition | Detritus fate | Other production | Fishery | Growth | Units

Figure 1. Basic input for Ecopath.

In addition, the program simulates a stable ecosystem also need to enter the feeding composition form, the information is estimated by the fish contents and the FishBase network, as shown in figure 2.

Input data

	Prey \ Predator	1	2	3	4
1	Omivorous				
2	Herbivorous		0.010		
3	Zooplankton	0.300	0.040	0.030	
4	Polychaetes	0.100			
5	Phytoplankton	0.100	0.450	0.470	0.280
6	Hydrophyte	0.100	0.120	0.200	0.120
7	Detritus	0.400	0.380	0.300	0.600
8	Import				
9	Sum	1.000	1.000	1.000	1.000

Basic input | Diet composition | Detritus fate | Other production

Figure 2. Prey /predato.

4. Preliminary results analysis

Preliminary analysis by software, the results shown in figure 3.

Parameter estimates

	Group name	Trophic level	Habitat area	Biomass in habitat area (Wet	Biomass (Wet weight (t/km?)	Prod./biom. (/year)	Cons./biom. (/year)	Ecotrophic efficiency	Production / consumption
1	Omivorous	2.41	1.000	40.000	40.000	0.300	1.200	0.000	0.250
2	Herbivorous	2.05	1.000	50.000	50.000	0.700	2.800	0.040	0.250
3	Zooplankton	2.03	1.000	1.400	1.400	20.000	62.000	0.807	0.323
4	Polychaetes	2.00	1.000	57.220	57.220	1.100	14.000	0.076	0.079
5	Phytoplankton	1.00	1.000	7.000	7.000	50.000	-	0.951	-
6	Hydrophyte	1.00	1.000	500.000	500.000	1.200	-	0.225	-
7	Detritus	1.00	1.000	10.000	10.000	-	-	0.718	-

Basic estimates | Key indices | Mortalities | Consumption | Respiration | Niche overlap | Electivity | Search rates | Fishery

Figure 3. Basic parameter estimates.

Trophic level column shows the average trophic level of each functional group in the figure, the highest class is omnivorous fish, its value is 2.41, it's very close to 2.38 which is checking from FishBase net, without calculating. The ecological efficiency (EE) calculated in the program is a reasonable value , because it means there is no natural death (starvation, aging and other factors of death), when EE is equal to 1.0. at the same time, the highest ecological efficiency of phytoplankton which value is 0.951 in line with Tongling [1] describing the general phytoplankton ecological efficiency is about 0.95.

5. Statistics of total parameters of the water ecosystem

The program statistics the parameters of the water ecosystem simulation of inland waterways, as shown in figure 4.

Summary Stats | Food Intake | Niche: Prey

Parameter	Value
Sum of all consumption:	1075.880
Sum of all exports:	227.238
Sum of all respiratory flows:	722.762
Sum of all flows into detritus:	806.326
Total system throughput:	2832.206
Sum of all production:	1087.942
The fishery has a 'mean trophic level':	-
Gross efficiency (catch/net p.p.):	-
Input total net primary production:	-
Calculated total net primary production:	950.00
Unaccounted primary production:	-
Total primary production/total respiration:	1.314
Net system production:	227.238
Total primary production/total biomass:	1.449
Total biomass/total throughput:	0.231
Total biomass (excluding detritus):	655.620
Total catches:	-
Connectance Index:	0.472
System Omnivory Index:	0.016

Figure 4. Summary stats.

The total net primary production of the whole water ecosystem is 950 t/km^2/year, and the sum of all respiratory flows is 722.762 t/km^2/year, the ratio of the two is 1.314.When the ecosystem reaches a mature state, total net primary production will be very close to or equal to the sum of all respiratory flows, so we can see the inland waterway ecosystem has reached a mature state.

6. Conclusion

We choose the measured data, synthesizes the domestic and foreign literature, constructing and simulating the food network energy transfer model of the water ecosystem in the inland waterway with the Ecopath software. Further explores the food source of fishery creatures in inland waterways and the transmission path of its energy in the food web by calculating the amount of energy transferring between the biological, biological and organic debris, and the estimating amount of biological respiration in the energy balance. The results shows that most of the material transport in the water ecosystem is transported to the upper layer by means of feeding, or the energy is consumed by respiration, and a small part of the material forms debris, the debris is returned to each biological group by feeding, and form the biogeochemical cycle. It's indicating that the ecosystem is equilibrium when the total net primary production is very close to the sum of all respiratory flows. The research method of this paper can be widely used to evaluate the stability of the ecosystem of the domestic river.

Acknowledgement

This work was supported by the Central Public Welfare Fund（Tks160209）.

References

[1] Tong L 1999 Ecopath model-a mass-balance modeling for ecosystem estimation *J. Marine Fisheries Research* **20(2)** pp 102-107

[2] Tong L, Tang Q S and Daniel P 2000 A preliminary approach on mass-balance ecopath model of the Bohai Sea *J. Chinese Journal of Applied Ecology* **11(3)** pp 435-440

[3] Froese R, Pauly D. Fishbase 2006: World Wide Web electronic publication[EB/OL].http:// www.fishbase.org

[4] Rodier, Aubert L B and Michel K 1997 Plankton biomass and production in an open atoll lagoon *J. Journal of Experimental Marine Biology and Ecology* **212(2)** pp 187-210

[5] Jeppessen E, J P Jensen and P Kristensen 2000 Fish manipulation as a lake restoration tool in shallow, eutrophic, temperate lakes 2: thresholds, long terms stability and conclusions *J. Hydrobiologia* **201** pp 219-227

[6] Pauly D 1998 Use of ecopath with ecosim to evaluate strategies for sustainable exploitation of multi-species resources *.Fisheries Centre Research Report* 6(2).University of British Columbia, Vancouver BC .Canada. p49

[7] Walters C J, Christensen V and Pauly D 1997 Structuring dynamic model of exploited ecosystems from trophic mass-balance assessments *J. Rev Fish Biol Fish* **7** pp 139-192

[8] Yang L L, Jiang Y Z and Yuan X W 2015 Evaluation on the ecosystem structure and function of Xiangshan Bay based on Ecopath model *J. Marine Fisheries* **37(5)** pp 399-408

[9] Odum E P 1969 The strategy of ecosystem development *J. Science* **164(3877)** pp 262-270

[10] Christensen V and Walters C J 2004 Ecopath with Ecosim: methods, capabilities and limitations *J. Ecological Modelling* **172(2-4)** pp 109-139

IOP Conf. Series: Earth and Environmental Science **133** (2018) 012027 doi:10.1088/1755-1315/133/1/012027

The construction strategy and measures for ecological analysis of China's ports

Lu Zhang, Junjie Zhao, Youping Shou, Ning Wang, Jianzhe Qiao1 and Mingjing Tian*

Abstract. The Port is an important transport hub, one of the important infrastructure of the national economy, the corresponding regional development has a great role in promoting.The construction of the port has been developed rapidly, but it has also brought about more serious environmental problems. Therefore, this paper discusses the nature of ecological ports, summarized the experience of ecological protection measures on the international advanced ports and port construction ecology. The central idea is to explore the construction of ecological port ideas and development countermeasures.

1. Introduction

The Port is an important transport hub, one of the important infrastructure of the national economy, the corresponding regional development has a great role in promoting. The construction of the port has been developed rapidly, but it has also brought about more serious environmental problems. Now, the port has been unable to meet the environmental requirements, therefore the port must be consider the transition. Whether the environment and the port can be coordinated has become a hot issue, therefore, the ecological port came into being.

2. Development Status of Foreign Ecological Port

Developed countries have put forward the construction of ecological ports, the environmental protection work to the port to the overall planning, design, construction and operation, the concept of sustainable development applied to the port construction, making the ecological port quite effective.[1]

New Jersey Port, New York, US. The construction of the ecological port of the New Jersey Port Authority is based on the relevant environmental management system of the perfect ecological port. First, through the external publicity and the internal staff training, the whole concept of sustainable development has been greatly increased and integrated into the whole port development. Followed by the expansion of high-speed railway and the improvement of port logistics and so on measures, greatly promoted the ecological port built in the development of such a set of sustainable development concept has been provided.

Tokyo Port, Japan. The port to the environment as a starting point, planning a meticulous on the port of the green park planning, green space planning line, line, surface, and connected to form a large network of green space system. Coastal and water landscape protection development as the focus of environmental construction, beach restoration has also been taken seriously, in addition to the port and environmental planning combined, the environment and port development combined, this series of measures makes the ecological port Of the built to be guaranteed.

Throughout the world some of the Western developed countries, ecological port construction has become increasingly mature, can be a good idea of sustainable development into the daily construction and movement, but also has its own management of the port regulations and policies. Such as the US federal government and state government environmental agencies have made clear environmental protection targets for the port to improve the target and to enforce the relevant measures. Regulatory authorities have also been strict inspection and review of port environmental protection, Environmental management and strengthening of these ports can do their own, It can be said that this environmental awareness has entered the entire operation process, in addition to the government and the community have the full support. It can be said that the Western developed countries of this series of related measures on the one hand by the development of modern port adaptation,

On the other hand also reflects the true meaning of the ecological port.[2]

3. Current Situation and Problems of Ecological Port Development in China

3.1. Current Situation of Ecological Port Development in China

Compared to the western developed countries, China's ecological port is really a starting stage, they are mainly in Tianjin, Shanghai, such a large port.

In 2005, The Shanghai Port began to carry out the green port construction planning related research, of which the ecological port as the core issue of research,and Explore some of the relevant environmental protection programs. Tianjin Port has done a lot of effort to optimize to optimize layout, specifically divided into the Beijiang and Nanjiang region. At the same time to promote the construction of cargo facility for green building in this area increased investment. Qingdao port adhere to science and technology, talent policy, ideas innovation, scientific and technological innovation, through the development of science and technology out of their own characteristics of sustainable development.

3.2. The Problems in the Development of Ecological Port in China

At this stage, China's Port from the second generation to the third generation of port transition period, Compared with the western countries, China's ecological port construction of the degree of attention there is still a certain gap. For a long time, the lack of environmental awareness, in the port production is not the pollution treatment facilities within the scope of consideration, The entire port construction without ecological port awareness, the overall still remain in the pollution control and port area of greening, so that the overall port of China is facing extremely serious environmental problems. Not only that, the academic aspects of the lack of ecological port planning and design research, there is no complete sustainable development and ecological construction of the port system. On the ecological port development, construction, planning and operation of the evaluation index is not enough in the science, in addition, the government departments are also lack of ecological port-related aspects of the construction of the regulations, etc., on the ecological green-related publicity is also less.

4. Suggestions and Measures for the Development of Ecological Port in China

4.1. Building a resource-saving port

From the port planning, layout to the construction, production, the full implementation of environmental protection measures, the implementation of the whole process of environmental supervision and management, to achieve the port of conservation development, clean development, safe development and sustainable development. To achieve the port economic, social and ecological benefits together to improve. For the relevant ports and enterprises of energy-saving emission reduction and related environmental protection standards, regulations, etc. to do a good job in the implementation of the work, the real eco-friendly, environmentally friendly port construction standards, complete laws and regulations, law enforcement supervision and management system effective.

4.2. To carry out ecological port planning research, the establishment of ecological demonstration port area

Planning is the port construction and development of the action guide, to carry out ecological port planning research work is the construction of China's ecological port of the most powerful support. In the investigation and evaluation of the ecological environment in the harbor area, the study of the environmental capacity of the harbor area, the environmental impact analysis of the port construction and the assessment of the value of the marine ecosystem services and the analysis of the ecological port construction mode and experience in foreign countries, Follow the different functional types and geographical environment of the port to formulate the corresponding work points and priorities. For the ports that have been built and planned, the corresponding ecological construction requirements and strategies should be formulated according to the different construction period and operation period and the port ecological function Planning and factor allocation principles.[3]

4.3. Scientific layout, optimization of port resource allocation

Contemporary high-speed development of science and technology at the same time more close to the relationship between human and nature, Therefore, it is necessary to do a good job in the requirements of the scientific development concept under the port resource recycling system, including the port planning, construction and production of the whole process will be closely linked with environmental protection projects. To further coordinate the relationship between port development and resource utilization, to carry out performance supervision and management of the whole process of port construction and operation, to solve the phenomenon of overcapacity caused by the irrational allocation of resources in China's port, so as to realize port resources Rationalize the configuration.

5. Conclusion

China's basic national policy is resource conservation and environmental protection, therefore, in the new era of green eco-friendly port is not only the development needs of the times, but also The rapid development of economic high-speed globalization today has enabled the port's core competitiveness to be enhanced and sustainable development to achieve the fundamental path in the port.

Reference

IOP Conf. Series: Earth and Environmental Science **133** (2018) 012027 doi:10.1088/1755-1315/133/1/012027

[1] Fangmin Zhao, Dazhi Wang, Xin Wei 2007 Study on Environmental Impact Assessment of Port Master Planning. *Environmental Science & Technology*. **30(7)**:64-66

[2] Bin Chen, Luping Zhang 2003 Practice of Strategic Environmental Assessment for Port Planning.Shanghai Environmental Secience,**22(12)**:1013-1016

[3] Honyun Du,Hongxing Shi 2004 Port Development and Environmental Protection. *China Transportation Review*, **(4)**:30-32.01

[4] Shu Rui Wang, Yuanyuan Wang 2013 Current Situation and Countermeasures of Green Ecological Port Construction in China. *China Academic Journal Electronic Publishing House*.**(213):**11-25

[5] Chaofeng Shao, Meiting Ju 2012 Analysis of Status-Que of Ecological Port Construction at Home and Abroad and Its Enlightenments. *China Academic Journal Electronic Publishing House*. **(178):**69-72

[6] Rong Yao, Xiaofeng Chen, Na Zhang 2003 Brief Analysis of Green Ecological Port. *Yangtze River*.**(5)：** 28-29

[7] Lei Liu,Minfeng Ma, Fan Yang 2017 Thoughts and Development Countermeasures of Some Port Planning in China. *ZHONGGUO SHUIYUN*.**(7):**21-24

[8] Kuishu Zhang,Ziai Lu 2010 Study on Ecological Evaluation Method of Port Environmental Resources. *JOURNAL OF JIANGSU UNIVERSITY OF SCIENCE AND TECHNOLOGY*. **(24):**179-183.

Research on the Development and Enlightenment of Urban Environmental Engineering

Mingjing Tian, Guanglou Li, Lu Zhang, Youping Shou, Yajuan Li, Wei Ye, and Jing Xu*
Key Laboratory of Environmental Protection in Water Transport Engineering Ministry of Communications, Tianjin Research Institute for Water Transport Engineering, Tianjin 300456, China.

Abstract. In recent years, under the promotion of reform and opening up, China's economic development has greatly accelerated, urbanization is also gradually accelerated. In the process of urbanization, there are many problems. The development of environmental engineering is one of the most important points. While building our living environment; we should also pay attention to the implementation of sustainable development strategies. First of all, This paper describes basic situation of environmental engineering, and finally provided some measures to promote the strengthening of China's environmental engineering

1. Introduction

With the acceleration of urbanization, there will be a large influx of people, which will lead to a variety of pollution problems and lead to the decline in urban environment. Although the construction of urbanization has been success, but the impact on the environment has become the next stage should focus on solving the problem. To solve this problem, environmental engineering is the first thing we have to consider now. To fully understand the environmental engineering, have to find out the difficulties and according to the problems of which to develop a solution. For China's urbanization to speed up the search for effective ways, this requires us to work together to exert wisdom, improve the environmental engineering problem, and provide better environmental conditions for our urban development.

2. Overview of environmental engineering

Environmental engineering which means is the science and technology of environmental protection. Specifically, in order to avoid environmental improvements in the process of environmental damage caused by the improvement measures. Environmental engineering covers a wide range of topics, including many disciplines, and the process of environmental engineering construction will be more complex, in order to fully grasp it more difficult.

IOP Conf. Series: Earth and Environmental Science **133** (2018) 012028 doi:10.1088/1755-1315/133/1/012028

Environmental engineering has only recently began to become popular, still in the development stage, it has been able to solve some basic problems, so that the environmental engineering continues to develop, will certainly China's environmental problems to provide a great helpful. The problems faced by environmental engineering are brought about by urban development. Therefore, it is only when the environmental project is the prerequisite for urban development so that our country can achieve better urban development without destroying the environment.

3. The status of environmental engineering

With the acceleration of urbanization, China's environment has also been seriously damaged. At this stage, China has been paying more attention to environmental issues. So at, environmental engineering construction will take as an important development goal.

But there are some problems, this paper briefly expounds the present situation of environmental engineering construction in our country which to find exist problems.

In China's environmental construction, already have some results, but the sewage problem is still one of the problems need to solved. There are two main types of urban sewage, the first is the sanitary wastewater, the second is industrial wastewater. At this stage of the environmental engineering in the handling of related issues, means and technology are relatively backward that cannot solve this problem well. Urban sewage system construction is not complete, industrial waste water contains toxic substances; These problems need to be solved. If these problems cannot be dealt with well, the quality of environmental engineering cannot be improved.

The municipal refuse is second problem, at this stage of the garbage disposal in most areas has been using the traditional way, there are two ways which are buried and burned. The city produced thousands of tons of garbage by each day, the use of such a way it is difficult to completely deal with it and also did not make recycling, therefore the resources that can be reused are wasted. And it incineration of air pollution, buried plastic products are difficult to be degraded, Such a deal has a serious environmental impact and requires significant cost.

In the process of urban environmental engineering construction, landscaping can reduce noise and improve air pollution, On the improvement of urban living environment has a very important impact. But at this stage of urban construction, the importance of the importance of landscaping and not so much attention. The construction of various commercial housing as a key project, no focus on environmental issues.

Now the layout of the city is too close, there is no land can be green. At this stage, China's urban greening area has a certain gap compared with the developed countries. China's own urban greening speed and urban environment compared to the construction of a lot of backward.

With the accelerated construction of urbanization, various factories gradually increased. The number of cars used is gradually increasing, the air caused serious pollution and there are noise pollution and other issues. Face these problems, China has introduced a certain management policy. To be completely resolved, it is necessary to strengthen the management system, so that it is more balanced with the urbanization.

4. Measures to strengthen environmental engineering

Firstly, in order to speed up the construction of environmental engineering, the first thing is strengthen the concept. First of all, let the city residents realize that environmental protection is the needed for everyone's attention. Government departments should increase publicity for the parties concerned, Play a supervisory and promotional role in related work. So that all residents have a certain understanding of this, city residents can provide their own opinions or suggestions for the construction of environmental engineering.

Secondly, to increase the intensity of the funds for the construction of environmental engineering, encourage all kinds of social funds and foreign capital to invest in urbanization. It is difficult to rely solely on the government's investment is to completely solve the problem of environmental construction, so the introduction of funds, it can be a good solution to this problem.

Thirdly, Promote recycling, in the incineration process, 30% of the wastes has been incineration, more than 60% of the waste directly discharged into the natural environment, these wastes have a bad impact on the environment. Therefore, you can use the waste screening process, on the one hand to play a role in protecting the environment, on the other hand can also save a lot of resources.

Fourthly, can be supervised by the news media, So that this work can be better completed. At this stage, the media can attract attention, need to guide the public can be aware of the importance of environmental protection, to form a sense of consciousness. Do not pay attention to the environmental protection of the enterprise announced, through the power of public opinion to make it improved, and in the relevant aspects of doing a good business to commend, make it the industry benchmark. And to give the public the opportunity to express their views so that everyone can express their own things, put forward the various ideas for environmental protection, to enhance public awareness of environmental protection.

5. Conclusion

The city in order to maintain the development, it needs from a long-term point of view, cannot be considered from the interests of the present stage. Pay attention to economic development, but also the environmental engineering problems as an important part of urban construction to complete, so as to maintain the harmonious and stable development of the city.

Although there are a lot of problems in China's environmental engineering, but cannot give up the development. All living in the city people need to make some contribution to environmental engineering. So that the environment in the development of the gradual improvement of our future generations to create a better living environment.

Acknowledgment

This work was supported by the Central Public Welfare Fund（Tks160209）.

Reference

[1] Guofeng Li, 2005, Environmental Engineering Geological Problems and Sustainable Development in Urban Construction of Tangshan City[J]. *Hebei Ligong Xueyuan Xuebao*.**(01)**:121-124

[2] Changtao Liu, Qingji Shen, 2005, Environmental Ethics and Its Enlightenment to China 's Urban Development[J]. *Modern Urban Research*. **(06)**: 57-62

[3] Qingyuan Jiang, Ji Zhu, 2001, Developing Environmental Economy and Building a New Type City - Inspiration from Dalian 's Economic[J].*JING JI SHI*.**(08)**.91

[4] Jowitt Paul, 2004, Sustainability and the formation of the civil engineer[J]. *Proceedings of The Institution of Civil Engineers-engineering Sustainability*. **(157).2**: 79-88

[5] Jianlei Niu, 2004, Some significant environmental issues in high-rise residential building design in urban areas[J]. *Energy and Buildings*. **(36).12**: 1259-1263

[6] Yedla Sudhakar, 2005, Modified landfill design for sustainable waste management [J]. *International Journal of Global Energy Issues*. **(23).1**: 93-105

[7] Tang Bosin, Siuwai Wong, Milton Chihong Lau, 2008, SOCIAL IMPACT ASSESSMENT AND PUBLIC PARTICIPATION IN CHINA: A CASE STUDY OF LAND REQUISITION IN GUANGZHOU [J]. *Environmental Impact Assessment Review*. **(28).1**: 57-72

[8] Wolch Jennifer, Jason Antony Byrne, Joshua P. Newell, 2014, Urban green space, public health, and environmental justice: The challenge of making cities “just green enough” [J]. *Landscape and Urban Planning*.**125**: 234-244

[9] Ohunene Omole Catherine, Ozoji Bernadette Ebele, 2014, Science Education and Sustainable Development in Nigeria[J]. *Education*. **(2).8**: 595-599

[10] Zhi Wang, 2017, The Aesthetic Function of Urban Environment Design[J]. *Heilongjiang Science and Technology Information*. **(02)**.235.

Strategic Environmental Assessment and Sustainable Development

Mingjing Tian, Qingjun Gao, Nan Wang, Xigen Yang, Xin Xu, and Lu Zhang*
Key Laboratory of Environmental Protection in Water Transport Engineering Ministry of Communications, Tianjin Research Institute for Water Transport Engineering, Tianjin 300456, China

Abstract. The development of social science and technology economy, the international community more and more attention to environmental and development issues. So the main goal pursued by people is not only to meet the needs of social and natural resources, while at the same time being able to protect the needs of future generations. This is the path of sustainable development. Therefore, this paper is a detailed study of strategic environmental assessment and sustainable development.

1. Introduction

In order to fully implement the sustainable development strategy, in this report, the specific content of environmental protection has been made clear, That is, to control the total amount of pollutant emissions, also Prevention and control of water pollution and control of key cities of air pollution[1].One way to reduce environmental pollution is environmental impact assessment, Its development trend is the strategic environmental assessment, which mainly from the macro and practical aspects of the implementation of the sustainable development strategy, and it has an important role in environmental protection[2]. Therefore, the specific study of strategic environmental assessment and sustainable development has important practical significance.

2. A Basic Discussion on Strategic Environmental Impact Assessment

2.1. Definition of SEA

Strategic EIA is focused on reflecting the "From The Control Of The Source To The Process" of the strategic thinking. Mainly the normative and integrated, planned and its alternative to the impact of the environment. In the decision-making of public responsibility for the evaluation of the results of the application, it is mainly about the relationship between environment and development decisions from policy and coordination aspects of its planning. It can be seen, high level is the main feature of SEA, and a single project EIA does not have this feature[3].

2.2. The Development of Strategic Environmental Impact Assessment

In the 1960s, because Western countries simply thought that social development was economic growth, this idea led to the full outbreak of environmental crisis in the 1980s[4].The crisis has had a negative impact on economic growth, political stability and social equity. Western countries from the "source of prevention and control" of the environmental protection instead of "end of governance" and "control of the production process" environmental thinking. The process of transforming thought is the process of sustainable development, and thus put forward the overall strategy of solving environmental problems. Since the 1990s, our country has fully realized the importance and urgency of the SEA work. While the SEEA's major policies and regulations are clearly displayed in the relevant documents[5].In addition, China's strategic environmental assessment in all kinds of development plans is published in September 2003, "the People's Republic of China Environmental Impact Assessment Law". This is the first time in the comprehensive decision-making in the form of legal clear environmental protection. Implementation of the planning and operation of the entire project in the process of implementing the impact assessment of environmental impact assessment, so the inevitable trend of its development is the implementation of strategic environmental impact assessment.

3. The Feasibility and Necessity of Implementing Strategic Environmental Impact Assessment

3.1. Feasibility of Implementing Strategic Environmental Impact Assessment

3.1.1. Research on the theory and method of strategic environmental impact assessment

The Strategic Environmental Assessment System was developed by the US Environmental Protection Agency in the 1970s.The main part of the system is the computer's operating model and database[6]. The use of economic activities and resources and the prediction of environmental pollution and the spatial information of the whole forecasting department are described on the basis of economic, population and energy demand forecast. This is the main reason for the development of this system.

In the late 1980s, China was aware of the importance of SEA. Compared with the Western countries, learn from the experience and method of the implementation of SEA, it has been late.[7]

Since 1995, China's strategic environmental assessment theory, methods and system construction process is based on China's actual national conditions. At the same time, the empirical SEA was tried.

In recent years, Chinese scholars mainly study the improvement of environmental impact and the development of strategic environmental impact assessment, so as to provide theoretical basis for the practice of strategic environmental impact assessment.

3.1.2. Establish and improve the system

The implementation of environmental and development of comprehensive decision-making content of the provisions, at the same time major economic and social development decision-making EIA work carried out[8].

The primary responsibility of the State Environmental Protection Administration is clearly defined.

The status of strategic environmental assessment to be clear, at the same time in the environmental impact assessment planning regional, watershed and sea development.

3.2. The Necessity of Implementing Strategic Environmental Impact Assessment

IOP Conf. Series: Earth and Environmental Science **133** (2018) 012029 doi:10.1088/1755-1315/133/1/012029

3.2.1. Environmental impact assessment to effectively improve the objective requirements is the strategic environmental impact assessment

At present, China's management environment is a basic system of strategic environmental assessment system. Its role in the control and reduction of environmental pollution and the role of ecological and environmental protection is irreplaceable. The environmental impact of the EPB on a number of projects is not considered too much. And the impact of these projects on the environment is much higher than the planning and development of regional resources and major infrastructure construction[9].

At the end of the decision chain is the construction project. So can only agree or reject the specific items in the project EIA. There is no guiding role in the direction of policy or planning, so the protection of the environment cannot be carried out from the root causes. Cannot solve the macro and indirect effects, which is the main reason for this phenomenon. The SEA is generated in such circumstances as to be able to make decisions on all projects in the construction project.

3.2.2. Scientific ecological compensation is an important basis for strategic environmental assessment

Strategic EIA can objectively evaluate the effectiveness of the ecological compensation policy, ecological compensation is generally trans-regional and watershed. So it includes the distribution of the subject and object. Thus creating new ecological problems in the change of interest pattern. However, the SEA can analyze the distribution of benefits between the ecological compensation subject and object, While the objective evaluation of changes in the pattern of interests, Which can be used to evaluate the ecological compensation itself.

The key factor of ecological compensation policy is the standard and principle of ecological compensation. Therefore, the main reason for the purpose of ecological compensation policy is to evaluate the standards and principles of provincial ecological compensation.

4. Measures for SEA Work

4.1. The development of SEA should be based on environmental capacity

One of the laws of sustainable development and is environmental capacity,

Which is mainly the country or region in a certain period of time the maximum load capacity to accommodate pollutants in the unit range. Environment and natural resources are the prerequisites for sustainable development and this premise meets the requirements of environmental capacity. At present our main goal is to make the environmental capacity to adapt to the development of society, the total amount of contaminants is controlled from the source[10].

Our resources cannot meet the needs of the population, there is a certain contradiction between economic development and protection of the environment. Therefore, in the corresponding policies and planning to develop at the same time, we must make the economic development and environmental development to adapt, also pay attention to the issue of environmental capacity.

4.2. Strategic environmental assessment of the entry point is the EIA planning

In theory, the first implementation of the strategic environmental assessment policy, followed by planning regional and industry EIA, and finally the implementation of construction projects in the EIA.

In the implementation of planning policies and regulations after the environmental impact assessment, scientific evaluation of resources and environmental capacity is the implementation of the entire project before the work.

However, China's EIA is only required for planning EIA, so the strategic environmental impact of the comprehensive entry point can only be planning EIA. The strategic environmental impact assessment is to promote the planning of environmental protection.

4.3. The establishment of a mechanism for public participation in SEA

The establishment of a mechanism for public participation in SEA. The quality and effectiveness of SEA can be improved, which has a significant impact on the improvement of decision-making and the implementation of scientific decision-making.

The current public participation in SEA has a significant role in the discussion of viable methods and procedures. Which can promote the development of China’s comprehensive strategic EIA work?

5. Conclusion

Thus, with the socio-economic development, environmental and development issues more and more prominent. China's environmental protection put forward new requirements, resource conservation is one of the basic national policy. At the same time to the development of circular economy, to protect the ecological environment, which can make the conservation of resources, and thus promote the coordinated development of social economy and the environment?

So that the environmental protection work in the EIA work to be strengthened, and the strategic objectives of sustainable development can be achieved.

Acknowledgment

This work was supported by the Central Public Welfare Fund（Tks160209）.

Reference

[1] Yue Pan, 2005, Strategic EIA and Sustainable Development [J].*Comparative Economic & Social System*.**(01).06**:11-14

[2] Kelly Andrew H., Tony Jackson, Peter Williams, 2012, Strategic environmental assessment: lessons for New South Wales, Australia, from Scottish practice [J]. *Impact Assessment and Project Appraisal*. **(30).2**: 75-84

[3] Noble Bram F., Kelechi Nwanekezie, 2017, Conceptualizing strategic environmental assessment: Principles, approaches and research directions [J]. *Environmental Impact Assessment Review*. **62**: 165-173

[4] Mariano Enzo Barberio, Daisy Aparecida Do Nascimento Rebelatto, 2014, Transformation of wealth produced into quality of life: analysis of the social efficiency of nation-states with the DEA's triple index approach [J]. *Journal of the Operational Research Society*. **(65).11**: 1664-1681

[5] Yangguang Tian, Yingzi Nie, 2006, Strengthen strategic environmental impact assessment, and promote sustainable development [A]. *Chinese Society For Environmental Sciences*, CSES. **(03)**.3

[6] Nuckols John R., Mary H. Ward, Lars Jarup, 2004, Using Geographic Information Systems for Exposure Assessment in Environmental Epidemiology Studies [J]. *Environmental Health Perspectives* **(112). 9**: 1007-1015

[7] Wei Cha, Liu wei, Qi Xu, 2007, Strategic EIA is a guarantee of sustainable development [N].CHINA CHINA ENVIRONMENT NEWS.**(05).11.**05001

[8] Gamboa Gonzalo, 2006, Social multi-criteria evaluation of different development scenarios of the Aysén region, Chile [J]. *Ecological Economics* **(59).1**: 157-170

[9] Lei Wang, 2005, Promote Strategic EIA to Promote Sustainable Development. [N]Chinese Township Enterprise Magazine. **(12).08.**31002

[10] Michaelidis Basile, Anastasia Spring, Hans O. Portner, 2007, Effects of long-term acclimation to environmental hypercapnia on extracellular acid–base status and metabolic capacity in Mediterranean fish Sparus aurata [J]. *Marine Biology* **(150).6**: 1417-1429

IOP Conf. Series: Earth and Environmental Science **133** (2018) 012030 doi:10.1088/1755-1315/133/1/012030

Optimizing the selection of small-town wastewater treatment processes

Jianping Huang* **and Siqi Zhang**
School of Environmental and Municipal Engineering, North China University of Water Resources and Electric Power, Zhengzhou 450046, China

Abstract. Municipal wastewater treatment is energy-intensive. This high energy consumption causes high sewage treatment plant operating costs and increases the energy burden. To mitigate the adverse impacts of China's development, sewage treatment plants should adopt effective energy-saving technologies. Artificial fortified natural water treatment and use of activated sludge and biofilm are all suitable technologies for small-town sewage treatment. This study features an analysis of the characteristics of small and medium-sized township sewage, an overview of current technologies, and a discussion of recent progress in sewage treatment. Based on this, an analysis of existing problems in municipal wastewater treatment is presented, and countermeasures to improve sewage treatment in small and medium-sized towns are proposed.

1. Introduction

At the 2015 United Nations General Assembly, the 2030 Agenda for Sustainable Development was adopted and 17 Sustainable Development Goals were established. Goal 6, "Water and sanitation for all and its sustainable management", is considered a necessary step towards ensuring sustainable development in the area of water quality improvement [1]. As such, China has implemented a stringent water resources management system. The process of water user participation should be promoted, and the appropriate technology, processes, and modes should be selected in consideration of prevailing opinions and local conditions. In 2016, water consumption per 10,000 yuan of Gross Domestic Product dropped 5.6%.

The development of new sewage treatment capacity has been exceeded by China's rapid industrial growth and urbanization. From 2009 to 2015, the number of urban sewage treatment plants in China increased from 1878 to 3542, a compound annual growth rate of 11.25%, with a corresponding daily sewage treatment capacity increase from 105 to 170 million tons/day. Although the number of sewage treatment facilities in China has increased substantially, there remains a big gap between sewage treatment capacity and the overall societal demand for improving the water environment. In particular, the sluggish development of sewerage networks has led to inadequate utilization of sewage treatment facilities in many cities [2]. Compared with larger cities, small and medium-sized cities have invested much less in sewage treatment, and most of the wastewater has been discharged directly into surface water bodies. Worse still, in rural areas, basic sewage collection and treatment infrastructure often is nonexistent.

Among the cities in China, large cities generally have abundant resources and specialized management systems, funds, and personnel [3]. Furthermore, environmental protection policies have typically been in place for a long time, sewage treatment facilities are relatively mature, and treatment standards are relatively high. Conversely, small and medium-sized cities are generally just now in the midst of the process of adequately treating sewage. Regarding treatment facilities, certain new construction and reconstruction inputs have been put forth in recent years. Due to the relative lag in development funds in county towns, there are problems such as inadequate sewage treatment infrastructure and low treatment standards for existing facilities. Thus, in consideration of the collective development potential of county towns, wastewater treatment facility construction and transformation are likely to soon reach a peak phase.

2. The status of urban sewage treatment technology

2.1. Water pollution situation

Urban sewage is an important source of water pollution, and the proper collection, treatment, and discharge of municipal sewage is a very important countermeasure to reduce and prevent water pollution [4]. Sewage treatment plants play a critical role in this process. Based on survey (MOHURC 2010), the mostly adopted treatment technologies in municipal WWTPs of China include oxidation ditch (30.5%), the anaerobic/anoxic/oxic (16.2%), conventional activated sludge (10.0%), anoxic/oxic (8.2%), and sequencing batch reactor (6.8%) [5]. Small towns using these processes often face high construction and operating costs, along with difficulty in ensuring adequate technical support and other issues. As such, small towns often cannot effectively build or operate normal sewage treatment plants.

The main problems facing small to medium-sized urban sewage treatment systems include: (1) low population, low water consumption, and small sewage treatment plants; (2) specific regional differences, affected by the rainy season and the amount of water when the coefficient of change is large, such that changes in sewage water output result in water quality changes; (3) low levels of economic development, low affordability, and few applicable technologies; (4) construction and operating costs are high relative to the populated served; and (5) maintenance and technical personnel management, including a serious lack of operations and management experience can pose additional challenges.

2.2. The characteristics of urban sewage technology

Currently in China, it is difficult to treat urban sewage, domestic sewage, and industrial wastewater separately. Therefore, industrial wastewater comprises a considerable part of urban sewage, and urban sewage water quality often has higher suspended solids and organic matter concentrations than would be expected if these wastes were handled separately. However, urban drainage systems are imperfect, so rainwater and groundwater infiltrate into the sewage pipe network, and the resultant concentration of organic matter CODcr is likely to 250 mg/L or less. Therefore, to meet China's urban sewage treatment needs, future systems must have a higher suspended solids and organic treatment efficiency, a better nitrogen and phosphorus removal capacity, a strong impact load capacity, lower infrastructure investment needs, the provision of infrastructure funds, and must be easier to operate and manage [6]. The existing biochemical treatment-based processes in large-scale urban sewage treatment plants have solved some of the problems of urban sewage treatment, but they are not well-suited to the characteristics of urban sewage.

The state has a plan to invest more funds for urban environmental management. An increase in small-town sewage treatment is an excellent opportunity for the development of the environmental protection industry. However, in light of the current status of urban sewage treatment processes, the development of the environmental protection industry has certain difficulties. Chiefly among them is that current sewage treatment technology cannot meet the requirements put forth by the current level of industrialization. Due to the limitations inherent in the available processes, the sewage retention time is too long, such that, to ensure sufficient pool capacity, steel and concrete structures must be used. As such, the sewage treatment plant construction period is lengthy and requires a long-term investment. Furthermore, it is difficult for the sewage treatment facilities to acquire the appropriate equipment, in complete sets, and especially when constructed in a serial fashion. Therefore, the relevant supplier industry is scattered and small, and competition is fierce. Based on similar experience from other industries, the environmental protection industry must fully develop, and this development must be led by a systematic production of the necessary equipment. Such an undertaking would enable the development of the entire industry. Therefore, a focus in the development of sewage treatment-related industries is the efficiency and miniaturization of treatment facilities.

3. Urban sewage treatment process analysis

At present, the sewage treatment technologies commonly used in small town systems include: artificial wetlands, Constructed Rapid Infiltration (CRI), Continuous-flow Intermission Biological Reactor (IBR), Alternated Internal Cyclic System (AICS), and Rotating Biological Contactor (RBC). A comparison and analysis of suitable processes for sewage treatment in small towns was performed (table 1).

Table 1. Comparison and analysis of suitable process for sewage treatment in small towns

Item	Technology	Infrastru-cture costs	Opera-ting Cost	Manage-ment difficulty	Occupied area	Sludge quantity	Treatme-nt effect	Environmen-tally sensitive area
Artificial reinforce-ment of natural water treatment	Artificial wetland	low	low	low	big	few	worse	Affected by the climate
	CRI process	lower	low	low	big	few	worse	
Activated sludge process	IBR process	medium	medium	big	medium	big	good	Suitable for low temperature areas
	AICS process	higher	high	high	medium	big	good	
Biofilm process	RBC process	medium	lower	low	medium	less	better	

3.1. Artificial wetlands

Constructed wetlands are ecosystems with high pollutant removal capacity that simulate natural wetlands, including water, plantings, and other aquatic organisms. Filtration is generally provided from the soil, sand, gravel, and other components. Some pollutants are filtered, while others provide a growth medium for plants and microorganisms. Plants are selected for their treatment performance, high survival rate, and strong anti-load ability, thus enabling them to improve water quality, secure the bed surface, prevent wetland blockage, and provide a good root zone environment for microorganisms. Microbes can remove some of the pollutants as the season and plant growth changes, and are thus an important part of constructed wetlands. This treatment process keeps infrastructure costs low, requires low energy consumption during operation, has low operating costs, and there are no complex mechanical, electrical, or automatic control equipment needs, and operations and management have low technical requirements [7]. The disadvantage of this approach is that it necessitates a large land area, plus the sewage treatment efficacy changes based on the season, as its resistance to bad weather is low.

3.2. Constructed Rapid Infiltration (CRI)

CRI treatment processes use the high permeability of natural river sand, add a certain amount of special filler as a filter for seepage after re-infiltration, and offer wet and dry alternate operating modes to remove water pollutants. Due to the growth of organisms and the sedimentation of suspended solids, excessive clogging of the surface of the percolation tank can be prevented, thus effectively restoring the permeability of the system. Furthermore, a redox (nitrification and denitrification) environment can be formed in the shallow section of the system, achieving the effect of removing pollutants [8]. The CRI process treatment efficacy is good and stable. Initial project investment and operating costs are lower, and system management is easy. There is less sludge production, practically eliminating sludge disposal problems, and thus also nearly eliminating the problem of secondary pollution. With CRI, it is easy to split the facility into modular construction systems, thus offering both scalability and flexibility.

3.3. Continuous-flow Intermission Biological Reactor (IBR)

IBR reaction tanks are divided into reaction zones (middle pool) and sedimentation zones (pool on both sides). The reaction zone is equipped with submersible pumps. Wave mass transfer devices and agitators reduce the blower room and aeration pipeline by jet aeration. The mixture containing activated sludge passes through the three-phase separator on both sides of the bottom of the tank for gas, solid, and liquid separation. Inclined pipe packing is installed in the settling zone, and the precipitated sludge can flow back to the reaction zone by itself to complete the inner cycle without power. Aged sludge is discharged through the mud tank bottom, and water is discharged from the top of the tank. Characteristics of the IBR process include less structural components and a smaller footprint. With changes in the influent sewage quality, quantity, temperature, and seasons, adjustments can be made to the reaction tank aeration, mixing, and precipitation cycles to achieve maximum simultaneous nitrogen and phosphorus removal at the lowest energy consumption and lowest operating costs [9].

3.4. Alternated Internal Cyclic System (AICS)

AICS systems combines SBRs with activated sludge and oxidation ditch technology and are one of the most widely used methods in cold areas. They include hydrolytic acidification, with anoxic, anaerobic, and aeration units for nitrogen and phosphorus removal. They consist of alternating long-term and short-term processes, and the operating cycle of the system is generally 4-8 h. For the long-term process, the system has two aeration pools that can be used as sedimentation tanks. One acts as the influent to the aeration tank, while the other as the effluent for the sedimentation tank. The two aeration tanks in the middle always maintain the aeration state during both long-term and short-term processes. For the short-term process stage, the pond that is preparing to enter the sedimentation stage must maintain a short aeration and 0.5-h static precipitation to meet the final water quality requirements. This also ensures that the initial effluent water quality meets the appropriate standards when entering the long-term process. AICS systems have the advantage of good nitrogen and phosphorus removal and good system stability in cold areas, as they are minimally affected by temperature [10]. However, they require a larger land area, higher investment costs, and higher maintenance costs.

3.5. Rotating Biological Contactor (RBC)

RBCs developed on the basis of biofilters are widely used in small-town sewage treatment facilities across China. Microorganisms in the wastewater attach to the turntable disk as biofilm [11]. About 40-45% of the disk surface is immersed in the sewage. With each rotation of the disk, the biofilm absorbs the oxygen in the air and the nutrients in the sewage for one absorption-decomposition cycle. As the disk keeps turning, the sewage is purified. The RBC process is adaptable to fluctuating water quality and quantity, consumes little energy, has lower operating costs, is easy to manage; and is adaptable to lower temperatures.

4. Conclusion and prospects

Process optimization in small towns should meet the following principles:

(1) Lower investment costs. Practitioners should try to adopt money and energy-saving processes and equipment, minimize the number of processing facilities, use appropriate treatment processes to reduce excess sludge emissions, and reduce the required dosages to reduce certain chemical costs.

(2) Lower operating costs. Practitioners should choose shorter processes, fewer operating units, easier maintenance, and fewer operators, in order to reduce ongoing operations and maintenance costs.

(3) Stable and reliable influent quality and quantity. For small-town sewage water quality, changes in the influent characteristics can be quite large. As such, practitioners should seek anti-impact loads, highly able to regulate the process, with mature and reliable processes, and, if necessary, should include some nitrogen and phosphorus removal capability to prevent eutrophication.

(4) Simple operation, easy management. Due to the often-limited expertise of managers of small town sewage treatment facilities, these facilities should be conducive to operations and management by the available management staff.

(5) Low environmental impact. The construction and operation of sewage treatment facilities in small cities can have an adverse impact on the surrounding environment. Especially true with constructed wetlands given their large area, certain impacts on the residents near where the facilities are located may exist, and may even harm the local environment and ecology.

Acknowledgments
The study was supported by Henan Province Science and Technology Major Project (No.161100310700).

References
[1] UNESCO (United Nations Educational, Scientific and Cultural Organization), 2015.Water for a sustainable world. The United Nations World Water Development Report 2015. Available from: http://unesdoc.unesco.org/images/0023/002318/231823E.pdf (accessed 27.09.2015, in Chinese).
[2] Sun, Y., Chen, Z., Wu, G., Wu, Q., Zhang, F., Niu, Z., Hu, H. Y., 2016. Characteristics of water quality of municipal wastewater treatment plants in China: implications for resources utilization and management. Journal of Cleaner Production, 131, 1-9.
[3] BWA, 2013. Water Resources Bulletin of Beijing 2012. BWA, Beijing, China.
[4] Bagatin, R., Klemeš, J., Reverberi, A.P., Huisingh, D., 2014. Conservation and improvements in water resource management: a global challenge. J. Clean. Prod. 77, 1–9.
[5] MOHURC, 2012. Management information system on wastewater treatment in 2012, Beijing, China (unpublished, in Chinese).
[6] Wang, T., Liu, S., Qian, X., Shimizu, T., Dente, S. M., Hashimoto, S., Nakajima, J., 2017. Assessment of the municipal water cycle in China. Science of the total Environment, 607-608, 761-770.
[7] Hussein, A., Scholz, M., 2017. Treatment of artificial wastewater containing two azo textile dyes by vertical-flow constructed wetlands[J]. Environmental Science and Pollution Research, 2017(7):1-20.
[8] Zhu, W. T., Sima, X. F., Yu, L. P., 2012. Optimizing operational parameters of new constructed rapid infiltration system in villages and towns wastewater treatment[J]. Chinese Journal of Environmental Engineering, 2012, 6(5):1459-1466.
[9] Wang, Y. C., 2016. Design of ibr process for wastewater treatment plant in a small town. China Water & Wastewater. 18:60-63.
[10] Cui, Z. F., Wang, K. J., Jia, L. M, et al. Application of Alternated Internal Cyclic System[J]. 2004. China Water & Wastewater, 20(9):56-58.
[11] Daudpoto, M. R., Talpur, G. H., Rind, U. A., et al. 2017. Response Surface Methodology for the Removal of Chemical Oxygen Demand through Rotating Biological Contactor[J]. Sindh University Research Journal-SURJ (Science Series). 49(3): 617-620.

IOP Conf. Series: Earth and Environmental Science **133** (2018) 012031 doi:10.1088/1755-1315/133/1/012031

Variations of PM2.5, PM10 mass concentration and health assessment in Islamabad, Pakistan

Tariq Memhood[1*], Z Tianle[1], I Ahmad[2], X Li[1], F Shen[1], W Akram[2] and L Dong [1]
[1]School of Space and Environment, Beihang University, Beijing, 100191, China,
[2] National Center for Physics, Islamabad, Pakistan

*Email: tariqslamat@gmail.com

Abstract. Sparse information appears in lack of awareness among the people regarding the linkage between particulate matter (PM) and mortality in Pakistan. The current study is aimed to investigate the seasonal mass concentration level of PM2.5 and PM10 in ambient air of Islamabad to assess the health risk of PM pollution. The sampling was carried out with two parallel medium volume air samplers on Whatman 47 mm quartz filter at a flow rate of 100L/min. Mass concentration was obtained by gravimetric analysis. A noticeable seasonal change in PM10 and PM2.5 mass concentration was observed. In case of PM2.5, the winter was a most polluted and spring was the cleanest season of 2017 in Islamabad with 69.97 and 40.44 μgm^{-3} mean concentration. Contrary, highest (152.42 μgm^{-3}) and lowest (74.90 μgm^{-3}) PM10 mass concentration was observed in autumn and summer respectively. Air Quality index level for PM2.5 and PM10 was remained moderated to unhealthy and good to sensitive respectively. Regarding health risk assessment, using national data for mortality rates, the excess mortality due to PM2.5 and PM10 exposure has been calculated and amounts to over 198 and 98 deaths annually for Islamabad. Comparatively estimated lifetime risk for PM2.5 ($1.16\times10\text{-}^{6}$) was observed higher than PM10 (7.32×10^{-8}).

1. Introduction

Hastily particulate matter (PM) pollution in the air has been evolved all around the world, predominantly in developing countries. World health organization [1] claimed that high exposure to the particulate matter has potential to reduce projected life expectancies by 1 year which increases up to 5.5 years [2].

Airborne particulate matter emerges from both natural (volcanic eruption, forest fire and weathering of parent material and sea salt) and anthropogenic sources including power plants, internal combustion engines as well as from thermo-degradation process. These particles are known to be air toxins and were the subject of WHO guidelines in 2008 when 24-h maximum thresholds for fine (PM2.5) and coarse (PM10) particulate matter were set at 25 and 50 μgm-3 respectively. The extent of toxicity and residence time of particulate matter in the atmosphere directly associated with their morphology, size and surface area [1]. The literature showed that long persistent in the air and deeper penetration inside lungs makes PM2.5 more toxic than PM10 which cause respiratory and cardiovascular diseases due to short-term exposure and mortality by long-term exposure respectively [3].

The World Health Organization (WHO) reported approximately 360,000 premature deaths in Asia each year due to urban air pollution [4]. Pakistan is the most urbanized region in Asia with approximately 35% population resides in cities and towns which is predicted to increase up to 60% by 2050 [5]. This appears in ever increasing air pollution extent in all major cities of Pakistan including Karachi, Lahore, Quetta, Peshawar, and Islamabad. Previous studies have reported high TSP concentrations in Lahore and Karachi compare to other most polluted sites worldwide. Karachi ranked first among the top 18 polluted cities in the world with average 668 μgm-3 PM2.5 concentrations [6]. In the same lines, Lahore was reported to be more polluted with average PM2.5 mass concentration than New York City, Seoul, and Hong Kong. In 2001, a joint venture of Pak-EPA and JICA [7] monitoring of air quality of three cities (Rawalpindi, Islamabad, and Lahore) reported higher PM10 concentration in the ambient air than WHO limits i.e. 520, 709 and 895 μgm-3 respectively. Hence the PM10 and PM2.5 mass concentration were very higher than the national environmental quality standards for ambient air with maximum thresholds for PM10 (annual, 120; 24-h, 150 μgm-3) and PM2.5 (annual, 15; 24-h, 35 μgm-3) respectively [4]. So far, Pakistan bears economic health burden equal to 1% of the gross domestic product due to PM which is responsible for 22000 premature deaths among adults and 700 deaths among children in Pakistan [8].

In spite of the fact, monitoring and management of PM are not among the top priority in Pakistan due to insufficient information, inadequate law practices, and poorly managed organizations. Resultantly, little sporadic updates and reports which illustrated the emerging and seriousness of PM pollution in Pakistan could be found in the literature [7]. There is no sufficient work has been done on PM mass concentration in Pakistan, especially on PM2.5 size friction of PM which appears in lack of public awareness and increasing rate of mortality and respiratory diseases in Pakistan.

For a better understanding and measuring the extent of current PM pollution in the ambient air of Islamabad, the present study has been designed to measure PM10 and PM2.5 mass concentration in Islamabad. Apparently, the previous studies showed a significant change in PM mass concentration with a change in the season [9]. Hereby, the results were fourthly optimized by evaluating seasonal trends in change of PM concentration. In addition, it is the first time to measure the PM2.5 and PM10 mass concentration simultaneously in Islamabad. The results of this study will be used as a baseline for future studies and will be a good addition to existing data of PM pollution in Pakistan. Furthermore, the air quality index of Islamabad during this period along with health consequence of particulate matter pollution is also discussed in this study.

2. Material and Methods

2.1. Sampling Site

The current study was set to find out the PM2.5 and PM10 concentration in the ambient air of Islamabad. Islamabad is a capital city of Pakistan, located at approximately 500 m above sea level (Latitude 33∘49'N; Longitude 72∘24'N) and is 14 km northeast of its twin city Rawalpindi. It is surrounded by Margalla Hills in its north, and plains of Punjab and water bodies on other sides. The city expands over 906 km2 with a current population of about two million. Islamabad has four distinct seasons; warm and rainy summer (June to August), followed by dry autumn (September to November), cold and dry winter (December to February) and spring (March to May) [10]. Higher precipitation was recorded for summer during the monsoon while other seasons, for the most part, remained dry and the predominant wind direction was north-east. The industrial sectors, I-9/10 and Kahuta Triangle to the south of I-10, have high vehicular traffic density which is the main source of city's air pollution arising from obnoxious gases and metal particulates. The key industries present in these sectors are steel mills, marble factories, flour mills, oil and ghee units, soap/chemical factories, ceramics, paints, pigments, pharmaceutical manufacturing plants, and several other industrial units.

The samples were fixed 20 meters above from ground level on the top roof of Experimental Physics Directorate building at the national center for physics, Islamabad. The sampling site represents the clean urban area which has no influence on any specific air pollution source. It was approximately 4 km away from Murree road, 4.5 km from Bharakhu (high traffic area) and 22 km away from I-10 (industrial area). Filter-based PM10 and PM2.5 samples were collected on Whatman 47 mm quartz filter papers. The sampling was carried out across four seasons with two paralleled medium volume air samplers (Laoying 2030 intelligent TSP sampler and Laoying 2034 Heavy Metal in Air Sampler). All samplings were operated at a medium rate of 100 L/min and lasted for 24 h for each sampler. Total 30 days representative samples (excluding rainy days) were collected from each season. Samples were collected through all seasons of the year, the number at each site appearing in Table 1.

Table 1. Sampling schedule for each season of 2017 in Islamabad.

Season	Sampling dates (2017)	Total Samples
Winter	11 January-16 February	30
Spring	07 March-10 April	30
Summer	13 June- 10 August	30
Autumn	21 August- 01 October	30

After sampling, each loaded filter was stored in a refrigerator below 0°C before gravimetric and chemical analysis.

2.2. Gravimetric Analysis

The PM2.5 and PM10 mass were determined by gravimetry. Gravimetric analysis of the filters was performed by weighing sample filters before and after sampling period. Quartz filter papers were equilibrated in a temperature and relative humidity controlled chamber for 24 hours and then weighted using an electronic balance (CPA-26P, Sartorius, German). The mass difference determined attributed to the total suspended particulate matter.

2.3. Meteorological information

Meteorological data was received from Pakistan meteorological department and online source (www.wunderground.com).

3. Results and Discussion

The sampling site represents a typical urban site which is just 3 km away from the Quaid-e-Azam University, a well-known site used in much previous air sampling research works [11]. A sum of 240 samples, 120 samples for each PM2.5 and PM10 was collected which was equally divided into 30 for every season. The rain affected days were not included in our sampling.

3.1. PM2.5

The seasonal variation in PM2.5 mass concentration i.e. winter (31.90-121.50 μgm^{-3}); spring (19.70-60.16 μgm^{-3}); summer (11.58-79.94 μgm^{-3}) and autumn (26.56-74.62 μgm^{-3}) was presented in figure 1. During the whole sampling period, PM2.5 values fluctuate from 11.58 to 121.50 μgm^{-3}. Figure 3 showed the seasonal average PM2.5 mass concentration. The maximum days of each season in Islamabad observed a higher level of PM2.5 than WHO permissible limit (of 25 μgm^{-3}), however highest PM2.5 concentration (149.80 μgm^{-3}) was observed in winter, on 23 February 2017.

The observed concentration of PM2.5 in winter was almost two times more than that of the summer seasons in Islamabad same as reported by Rasheed et al. [6]. They observed a higher level of PM2.5 in winter with the highest hourly average concentrations as 303 μgm-3 (December 2007); 495.0 μgm-3 (November 2008); 259.8 μgm-3 (September 2009); 456.0 μgm-3 (October 2010) and 379.0 μgm-3 in January 2011 respectively. PM2.5 in winter with the highest hourly average concentrations as 303 μgm-3 during December 2007, 495.0 μgm-3 during November 2008, 259.8 μgm-3 during September 2009, 456.0 μgm-3 during October 2010 and 379.0 μgm-3 during January 2011. In Islamabad, a sporadic information available about the PM concentration especially PM2.5 which addressed average annual PM2.5 mass concentrations in the range of 81.1 ± 48.4 μgm-3, 93.0 ± 49.9 μgm-3, 47.8 ± 33.2

µgm-3, 79.0 ± 49.2 µgm-3 and 66.1 ± 52.1 µgm-3 during 2007 to 2011 respectively. Comparatively higher PM 2.5 mass concentration in winter could be due to the burning of agricultural residue in the surrounding areas during these months [12]. Mansha et al. [13] have also reported higher PM2.5 mass concentrations during fall and winter seasons as compared to summer and forest fire in Margalla Hills [6]. Hence PM levels in Islamabad were on the same order of magnitude as previous studies in this capital city [6, 7, 12, 13] and other urban locations in Pakistan [7, 14, 15], but were significantly higher than in less-populated Asian locations [4].

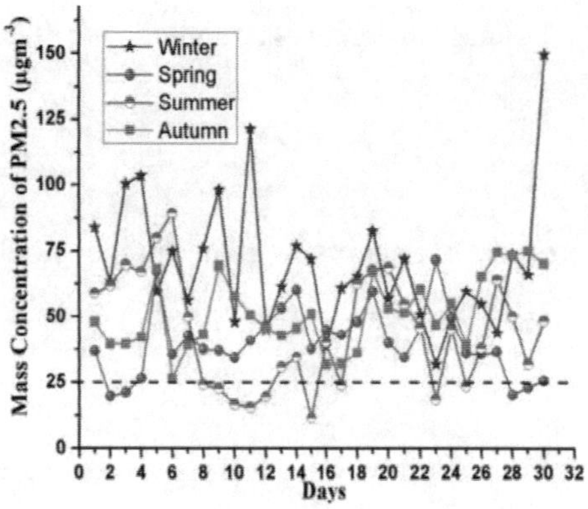

Figure 1. 30 days PM2.5 mass concentration

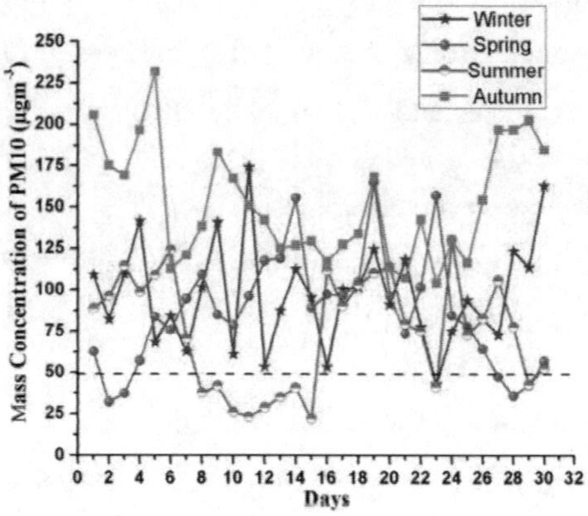

Figure 2. 30 days PM10 mass concentration

3.2. PM10

In 2017, a big variation (21.9-231 µgm^{-3}) in the mass concentration of PM10 was observed in Islamabad. However autumn and winter were a more polluted season with maximum days having a 24-h PM10 mass concentration above than permissible limit (150 µgm^{-3}) set by World Health Organization (WHO) and Pakistan National Environmental Quality Standards (Pak-NEQS). Maximum concentration of PM10 was found in autumn (104-231 µgm^{-3}) followed by winter (42.9-163 µgm^{-3}) spring (32-155 µgm^{-3}) and summer (21.9-130 µgm^{-3}) (figure 2).

The average PM10 concentrations were found to follow the decreasing order: autumn > winter > spring > summer; having 152.4, 97.22, 86.52 and 40.45 µgm-3 respectively (figure 3). Current results are in good agreement with the finding of Sha et al., 2010 in Islamabad [16]. They found a higher level of TSP (198 µgm-3) in autumn compared to the levels of TSP 138,148,190 µgm-3 in winter, summer and spring respectively. However, the observed PM10 concentrations for the year of 2017 in Islamabad were comparatively little higher than 2016 reported by Bulbul et al. [17]. They addressed the concentration of Islamabad in the winter, 2016 were in the range of 123-202 µgm-3 with an overall mean concentration of 177 µgm-3. Contrary, less pollution in spring and summer may be attributed to higher precipitation in Islamabad during monsoon (July-August) and in cold winter (January-February), the atmosphere was cleaned via washing out processes [18].

3.3. Meteorological condition during study period

A noticeable change in the PM concentration with a change in season is mainly attributed to a change in meteorological conditions [9]. The stable atmospheric condition with low temperate and high humidity facilitates the accumulation PM [18]. The higher values of PM10 and PM2.5 in winter and autumn in this studies may be directly influenced by the meteorological conditions as the lowest mean temperature 14.5°C with 44.5% humidity was observed in winter whereas, autumn has the highest humidity 51.5% with mean 37.6°C temperature. Although autumn observed higher temperature, it was well compensated by frequent rainfall [19], as it received higher average precipitation (126mm) compared to winter (93 mm), summer (88.5 mm) and spring (58mm) respectively. Additionally higher average wind velocity in summer (14.5 kmh^{-1}) and spring (14 kmh^{-1}) dilute air pollution more compare to autumn and winter with 12 and 12.5 kmh^{-1} respectively.

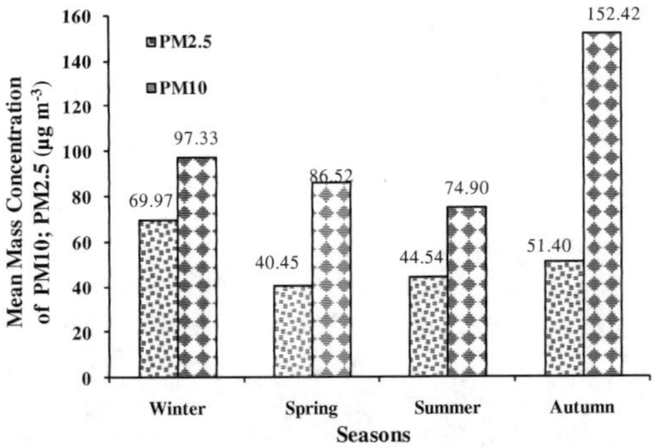

Figure 3. Average mass concentration of PM2.5 and PM10 in four seasons.

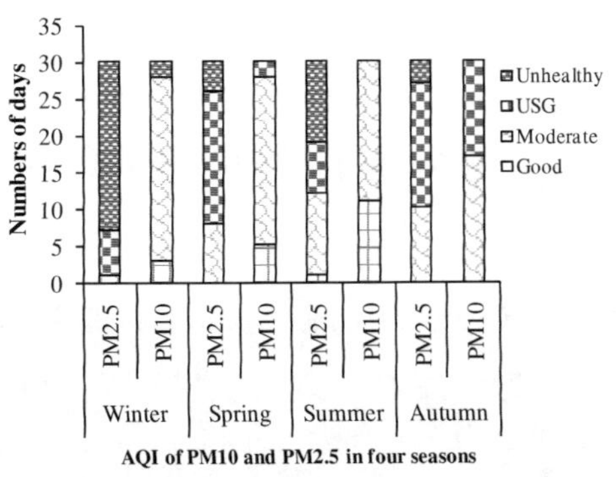

Figure 4. No. of days of each season with respect to specific AQI category.

3.4. Air quality index

Air quality index (AQI) is a relevant term used to describe the degree of the health risk of the air pollution and indicates its potential to affect the health. It is divided into 6 groups ranging from 0 to 500: good (0-50); moderate (51-100); unhealthy for sensitive individuals (USG) (101-150); unhealthy (151-200); very unhealthy (201-300) and hazardous (301-500) [20]. In Pakistan, the AQI related to

PM pollutants proposed by US-EPA (US-EPA, 2012) is being used to describe the air quality status. The AQI was calculated by relating PM mass concentrations to the relevant standards (US-EPA) [21].

Table 2. Air quality standards (annual mean mass concentration μgm^{-3}).

Pollutant	Time	WHO (2006) [22]	USEPA (2016) [20]	NEQSAA (Pak) 2010 [4]
PM2.5	(24-hours)	25	35	25
	(annual)	10	12	15
PM10	(24-hours)	50	150	150
	(annual)	20	54	120

The results of the current study showed that AQI of Islamabad remained confined from good to unhealthy (figure 5). The AQI for PM2.5 remained higher compare to PM10 throughout the sampling period. In case of PM2.5 AQI, winter was a most polluted season of 2017 in Islamabad as it contains 23 and 6 days with unhealthy and unhealthy levels for sensitive group correspondingly. However, autumn showed moderated to sensitive AQI for PM2.5 with 10 and 17 days respectively. On the other hand, AQI for PM10 remained comparatively well with 25, 23, 19 and 17 days of moderated pollution level in winter, spring, summer, and autumn respectively. The order of AQI of the current study was very similar to that found in Henan, China i.e. winter > spring > autumn > summer [23]. Summer stood most clean with 11 days of clean values of AQI. Results showed that AQI behaves periodically, peaking in the winter, same higher values of AQI for the winter was found in Xian, China. The numbers of days that are light, moderately, heavily and severely polluted air in the year 2016 are 71, 34, 32 and 8. These are all higher than those (66, 20, 16, 1) in the year 2015 in Xian [24]. However the overall AQI was observer lower in Islamabad (highest, 185) compare to the other big cities like Pakistan (Faisalabad, 909-715) [21]; India (Delhi, 336) [25]; Iran (Sistan, 316) [26].

3.5. Health risk assessment

Comparative finding from the observed mean concentrations of PM10 and PM2.5 for all four season (Fig.3) and the reference and standard concentrations (Table 2) showed higher values than the recommended values of the WHO and EU requirements. The method applied to calculate health risk was taken from the study conducted by Harrison et al. [3]. The lifetime risk can be calculated by use of equation no. 1;

$$\mathrm{LTR} = \frac{(\beta/C_{obs})}{L_e} \tag{1}$$

Where; LTR is Lifetime risk; β is the coefficient of giving pollutant; Cobs is an observed concentration of giving pollutant and Le is the life expectancy of exposed group.

Table 3. Health risk associated with the mean exposures.

Pollutant	Mean concentration	Coefficient	Lifetime risk	Endpoint
PM2.5	51.6	4%/10	1.16×10^{-6}	Mortality in all cases
PM10	102.8	0.5%/10	7.32×10^{-8}	Mortality in all cases

For all-cause mortality, PM10 and PM2.5, a coefficient has been obtained from WHO (2006) and rather than the usual mortality burden calculation, an incremental risk has been estimated for the mean concentration exposure assuming a mean life expectancy of 66.4 years [27]. Results showed high risk with the PM2.5 (1.16×10^{-6}compare) to PM10 (7.32×10^{-8}) in Islamabad. The risk associated with PM exposure subsequently increases due to chemical carcinogens attached to PM [6]. The calculated PM2.5 associated lifetime risk in Islamabad is lower than the finding of Harrison et al. in Saudi Arabia (Jeddah,1. 2×10^{-3}) [3].

3.6. Premature mortality due to PM2.5 and PM10 exposure

According to world population review, the current population of Islamabad is 1,502,000 persons (males and females). The data of population was compared with the average PM10 and PM2.5 mass concentration of four seasons obtained from current study to estimate premature mortality in Islamabad (Table 4).

Table 4. Premature mortality due to PM2.5 and PM10 exposure.

Pollutant	Observed values (mean)	Population in thousands [28]	Crud death rate [29]	Total Premature death
PM2.5	51.6	150.2000	6.4/1000	198
PM10	102.8			98

The premature mortality was simply calculated through the method (equation no. 2) used by Harrison et al. [3].

$$\mathrm{PMR} = \frac{T_p \times CDR}{C_t} \times C_{obs} \qquad (2)$$

Where; PMR is Premature mortality rate; Tp is total population exposed to a pollutant, CDR is crude death rate, Ct is toxic mean concentration (daily*annual) of pollutant defined by WHO and Cobs is observed the concentration of specific pollutant.

The method used to determine the level of premature mortality; the numbers are directly related to the number of people exposed and exposure concentration of PM10 and PM2.5. Thereby, the number of premature mortality will be more in a larger city with a similar concentration of PM, in the case of smaller cities, the number will enhance directly with increasing level of PM concentration. The present study showed the average four seasons values of PM2.5 and PM10 were 51.6 and 102.8 μgm^{-3} respectively. These values are within the range of previously reported average annual PM2.5 mass concentration values (81.2 ± 47.4 μgm^{-3}) by Rasheed et al. for the duration of 2007-2011 [6].

According to reported studies, 865,000 premature deaths are caused every year by air pollution and about 60% of these deaths found occurred in Asia [1].

4. Conclusion

The current study showed higher particulate matter pollution in Islamabad during the four seasons of 2017. The maximum days from all season have higher PM2.5 and PM10 values than the NEQSAA and WHO. The positive trend has been observed between PM10 and PM2.5 mass concentration; however PM10 values remain higher than PM2.5 in all sampling days. In case of AQI, days with PM2.5 pollution was more compared to PM10. In general, AQI results of PM2.5 from this study addressed winter as the most polluted season in Islamabad for the year of 2017 followed by spring, autumn and summer descendingly. Overall, winter and autumn were observed to be more polluted seasons followed by spring and summer respectively. This study only carries the mass concentration and health consequence with public exposure to particulate matter exposure. The health assessment for the mean mass concentration level of PM2.5 and PM10 observed in this study showed significant incensement in premature mortality and associated lifetime risk in Islamabad. However mass concentration alone is only a part of the threat to life but other chemical species also enhance its effect on life. Thereby, a detailed study has been designed including chemical characterization and source identification. The results of the further investigation will be discussed in future.

Acknowledgement

The Authors are thankful to Beihang University Beijing, China and National Center for Physics (NCP) Islamabad, and Pakistan Meteorological Department (PMD) Islamabad, Pakistan. Moreover, Mr. Tariq Mehmood is also thankful to Chinese Government and Chinese Scholarship Council for providing him P.hD. scholarship.

References

[1] World Health Organization 2003 Health aspects of air pollution with particulate matter, ozone and nitrogen dioxide: report on a WHO working group, Bonn, Germany 13-15 January 2003

[2] Chen H, Burnett R T, Kwong J C, Villeneuve P J, Goldberg M S, Brook R D, van Donkelaar A, Jerrett M, Martin R V and Brook J R 2013 Risk of incident diabetes in relation to long-term exposure to fine particulate matter in Ontario, Canada. In: Environmental Health Perspectives (Online), p 804

[3] Harrison R M, Bousiotis D, Mohorjy A, Alkhalaf A, Shamy M, Alghamdi M, Khoder M and Costa M 2017 Health risk associated with airborne particulate matter and its components in Jeddah, Saudi Arabia *Sci. Total Environ.* **590** 531-9

[4] NEQSAA 2010 National Environmnetal Quality Standards for Ambient Air. The Gazet of Pakistan. Environment Protection Department. The Ministry of Environment

[5] UNPD 2017 World Population Prospects: The 2016 Revision – Key Findings and Advance Tables" (PDF)

[6] Rasheed A, Aneja V P, Aiyyer A and Rafique U 2015 Measurement and analysis of fine particulate matter (PM2. 5) in urban areas of Pakistan Aerosol *Air Qual. Res* **15** 426-39

[7] Pak-EPA/JICA 2001 Three cities investigation of air and water quality (Lahore, Rawalpindi & Islamabad). Pakistan Environmental Protection Agency/Japan International Cooperation Agency

[8] World Bank 2006 Pakistan: Strategic country environment assessment, vol. II, report No. 36946 PK. World Bank, Washington

[9] Zikova N, Wang Y, Yang F, Li X, Tian M and Hopke P K 2016 On the source contribution to Beijing PM2.5 concentrations *Atmospheric Environ.* **134** 84-95

[10] Khan F K 1991 A geography of Pakistan: environment, people and economy: Oxford University Press)

[11] Ahmed F, Shah M H and Shaheen N 2014 Diurnal and nocturnal variations of trace metals in urban atmospheric particulate matter from Islamabad, Pakistan *Environ. Earth Sci.* **71** 817-26

[12] SUPARCO 2009 Space Research in Pakistan 2008-2009: National Report to 38th COSPAR Scientific Assembly Bremen, Germany, Space and Upper Atmosphere Research Commission (SUPARCO) Headquarters, Pakistan

[13] Mansha M, Ghauri B, Rahman S and Amman A 2012 Characterization and source apportionment of ambient air particulate matter (PM2.5) in Karachi *Sci.Total Environ.* **425** 176-83

[14] Biswas K F, Ghauri B M and Husain L 2008 Gaseous and aerosol pollutants during fog and clear episodes in South Asian urban atmosphere *Atmospheric Environ.* **42** 7775-85

[15] Gurjar B R, Butler T M, Lawrence M G and Lelieveld J 2008 Evaluation of emissions and air quality in megacities *Atmospheric Environ.* **42** 1593-606

[16] Shah M H and Shaheen N 2010 Seasonal behaviours in elemental composition of atmospheric aerosols collected in Islamabad, Pakistan *Atmospheric Res.* **95** 210-23

[17] Bulbul G, Shahid I, Chishtie F, Shahid M Z, Hundal R A, Zahra F and Shahzad M I 2017 PM10 Sampling and AOD Trends during 2016 Winter Fog Season in the Islamabad Region Aerosol & Air Quality Research

[18] Cao J J, Lee S C, Chow J C, Watson J G, Ho K F, Zhang R J, Jin Z D, Shen Z X, Chen G C and Kang Y M 2007 Spatial and seasonal distributions of carbonaceous aerosols over China *J. Geophysical Res.: Atmospheres* 112

[19] Shah M H, Shaheen N and Nazir R 2012 Assessment of the trace elements level in urban atmospheric particulate matter and source apportionment in Islamabad, Pakistan *Atmospheric Poll. Res.* **3** 39-45

[20] USEPA 2016 Technical Assistance Document for the Reporting of Daily Air Quality-the Air Quality Index (AQI): US Environmental Protection Agency. In: Office of Air Quality Planning and Standards, ed D Mintz

[21] Javed W, Murtaza G, Ahmad H R, Iqbal and M. M 2014 A preliminary assessment of air quality index (AQI) along a busy road in Faisalabad metropolitan, Pakistan 5

[22] World Health Organization 2006 Air quality guidelines: global update 2005: particulate matter, ozone, nitrogen dioxide, and sulfur dioxide: World Health Organization)

[23] Shen F, Ge X, Hu J, Nie D, Tian L and Chen M 2017 Air pollution characteristics and health risks in Henan Province, China *Environ. Res.* **156** 625-34

[24] Tang Tang S, Yan Q, Shi W, Wang X, Sun X, Yu P, Wu J and Xiao Y 2018 Measuring the impact of air pollution on respiratory infection risk in China *Environ. Poll.* **232** 477-86

[25] Kumar A and Goyal P 2011 Forecasting of daily air quality index in Delhi Sci. *Total Environ.* **409** 5517-23

[26] Rashki A, Kaskaoutis D G, Rautenbach C J d, Eriksson P G, Qiang M and Gupta P 2012 Dust storms and their horizontal dust loading in the Sistan region, Iran *Aeolian Res.* **5** 51-62

[27] World Halth Rankings 2017 world life expectancy

[28] Review W P 2017 Islamabad Population 2017

[29] World Life Expectancy 2017 WORLDHEALTHRANKINGS

IOP Conf. Series: Earth and Environmental Science **133** (2018) 012032 doi:10.1088/1755-1315/133/1/012032

Research for Foreign Advanced Ports for Protection and Development Experience

Lu Zhang, Mingjing Tian, Junjie Zhao, Youping Shou, Ning Wang, Jianzhe Qiao and Guanglou Li*

Key Laboratory of Environmental Protection in Water Transport Engineering Ministry of Communications,Tianjin Research Institute for Water Transport Engineering, Tianjin 300456, China

Abstract. Into the twenty-first century, the process globalization of economic and trade is getting faster and faster, As of 2014, China's annual port trading amount topped the world， But in the port of environmental protection sustainable development approach with foreign advanced port environmental management concept has a big gap. Combined with the present situation of modern ports in China. Drawing lessons from foreign advanced environmental protection idea of port, in order to promote the protection of port environment in our country. The experience of protection and development of foreign advanced port environment will be discussed and discussed.

1. Introduction

In the concept of advanced foreign port environmental protection, primarily, need to understand the environmental significance of the port. The modern international port is not only an economic hub. It is a bridge for international trade exchanges, also a sensitive area for environmental protection. The choice of port construction is in the open sea and deep sea land boundary. Through port construction, the ability of self-purification of sea water has dropped greatly in the harbor area, and it is a weak area of environmental protection. Bearing the mainstay of national economic development, Economic activity cannot avoid human high frequency activities. In the trade activities of port construction, the impact on the natural environment cannot be avoided. Therefore, the sensitive area of environmental protection is the focus of attention both at home and abroad.

Developed countries attach great importance to the harmonious development of man and nature. In the process of project construction, the idea of clinging to the nature, Avoid a wide range of changes in the natural environment to maintain the balance of the ecological environment.

In the construction of the project, the design includes the environmental protection design plan. The use of efficient management programme and measures to minimize environmental damage. The following that we summarize the management policies of advanced port environmental protection in

IOP Conf. Series: Earth and Environmental Science **133** (2018) 012032 doi:10.1088/1755-1315/133/1/012032

foreign countries.

1.1. Conforming to the design philosophy the nature

In the concept of advanced port environmental protection, Construction concept aimed at eliminating economic benefits. The idea of "harmony between man and nature" goes deep into every design link of an engineering project. In the construction design of the project, the fully considering the dependence of the first biological group on the original natural environment. Find out the identify the weakest link in the chain, Simulation is done before the construction of the project, A chain reaction caused by the destruction of the weak point.

To reduce the natural environment for the purpose, carry out industrial design and optimize practical materials for engineering construction, Avoid the environmental pollution of toxic substances, Cause a large area of death of the bio group. Base on improve the integrity of the food chain, try to maintain the original ecological environment. By introducing new varieties of aquatic plants, To improve the ecological structure of the ecological environment after the construction of the port.

In the port improvement project used in underwater construction form of trapezoidal multiple section or change the direction of flow of water under the current port. Underwater foundations with different shapes are adopted in underwater engineering, Scatter the base of the uniform character, different structural forms are applied in different regions, For example, in microbial enrichment areas, use square raw materials, slow down the impact of the sea.

In the biota complex area, porous and undulating materials are used for construction, Conducive to microbial attachment and the breeding and foraging of fish stocks. The construction of ecological rocks, the ability to artificially apply appropriate agents, promote the growth and reproduction of benthos， ensure that the diversity of ecosystems is not destroyed, for the survival of marine organisms to create a good growth environment.

At the port of Hemingway in Russia, The construction of ecological Great Barrier Reef, an irregular porous box is put into the corresponding deep water area, method for building ecological wave elimination block, promote the restoration of biological population after construction, ensure that the original ecological environment is not destroyed, it has made great contributions in the field of environmental protection in advanced ports.

1.2. Eliminate sources of pollution

International advanced ports focus on the use of natural resources, Marine ports include tidal energy, sea land, wind energy, solar energy and other pollution-free clean energy. On the utilization of clean energy, modernization is not effective. Make use of the most common Wind Energy. Because the specific heat of land is different from that of ocean， there is a steady flow of wind throughout the day and at night, the use of wind power system to provide lighting, electricity and daily life staff.

In port trade, many diesel powered operating equipment is being replaced by progressively more power driven ones, to eliminate the source of direct pollution, the government put more manpower and material resources into the port green energy.

2. Main methods of environmental protection and management in foreign advanced ports

2.1. Improve the utilization of materials

At the moment of port construction, a great deal of earthwork, A great deal of earth produced during this period can be fully utilized. Hong Kong is a model of success in China, Hong Kong International Airport expansion is the use of engineering construction in the earthwork reclamation way, Improved utilization rate. The effective use of materials is an effective measure to protect the environment, therefore make waste profitable. The waste generated in port trade plays an effective role in protecting the environment effectively in the process of returning to the project.

2.2. Improve the environmental management system

In the environmental protection management system, implement comprehensive monitoring measures, subarea management, implement regional responsibility system, strengthening environmental management, discover pollution sources in time, take corresponding measures to ensure that the environment is effectively protected.

2.3. Adjusting the key functions of environmental in the port

With the development of society, frequent economic activity has damaged the living conditions of the surrounding population. Adjust the number and type of imports and exports to reduce the impact on the surrounding environment.

3. Conclusion

There is a gap between the results of environmental protection and the advanced port environmental protection in developed countries.

Combined with the actual conditions of China's ports, learn from foreign advanced port environmental protection concepts and methods, promoting the protection of port environment in China.

Acknowledgment

This work was supported by the Central Public Welfare Fund（Tks160209）.

Reference

[1] Wei Zhao, Yunbiao Li, 2010, Harbor environmental protection countermeasure, tentative plan[J]. *ZHONGGUO SHUIYUN.* (06)

[2] Fangmin Zhao, Xujing Bi, Ying Liu, Yuxing Wang, 2009, Comparison of environmental protection between Chinese ports and Canadian ports[J]. *ZHONGGUO SHUIYUN.* (02)

[3] Qiong Zhao, Meiting Ju, Shuxue Chen, Xinyu Bei, 2009, Study on port environmental management system[J]. *ZHONGGUO SHUIYUN.* (04)

[4] Penghua Wu, 2009, Exploration of green ecological port construction [J]. *MARINE ENVIRONMENTAL SCIENCE.* (03)

[5] Libin Fan, 2009, Development and reference of foreign ports [J]. *ZHU JIANG SHUI YUN.*(09)

[6] Yayi Xiao, 2016, The Research on Liquidity Risk Management of China's Commercial Banks [J]. *Journal of Systems and Software.* **(04).03**: 251-259

[7] Gu Yan, Qingjun Meng, 2016, Problems and Countermeasures in the Implementation of Sustainable Development Strategy for Environmental Protection Enterprises in the New Period[J]. *Journal of Systems and Software.* **(04).05**: 243-247

[8] Akca Selen Ozakar, 2016, The effect of Foreign Body Aspiration training on the knowledge level of pupils[J]. *Revista Brasileira De Otorrinolaringologia*. **(82).4**: 408-415

[9] Sovacool Benjamin K., Morgan Bazilian, Michael Toman, 2016, Paradigms and poverty in global energy policy: research needs for achieving universal energy access[J]. *Environmental Research Letters*, **(11).6**.

[10] Lam Jasmine Siu Lee, Jonatan Lassa, 2017, Risk assessment framework for exposure of cargo and ports to natural hazards and climate extremes[J]. *Maritime Policy & Management*. **(44).1**: 1-15

[11] Mirzabeiki Vahid, Violeta Roso, Per Sjoholm, 2016, Collaborative tracking and tracing applied on dry ports[J]. *International Journal of Logistics Systems and Management*. **(25).3**: 425-440

[12] Pettigrew Luisa M., Inke Mathauer, 2016, Voluntary Health Insurance expenditure in low- and middle-income countries: Exploring trends during 1995-2012 and policy implications for progress towards universal health coverage[J]. *International Journal for Equity in Health*. **(15).1**: 67-67

[13] Freeman Nick J., 2014, Navigating the Economic Reform Process[J]. *Southeast Asian Affairs*. **(1)**: 224-240

[14] Jiaoe Wang, Yang Cheng, Huihui Mo, 2014, The Spatio-Temporal Distribution and Development Modes of Border Ports in China[J]. *Sustainability*, **(6).10**: 1-18

[15] Yang Dong, Anthony T H Chin, Shun Chen, 2014, Impact of politics, economic events and port policies on the evolution of maritime traffic in Chinese ports[J]. *Maritime Policy & Management*. **(41).4**: 346-366

[16] Rong Yao, Xiaofeng Chen, Na Zhang, 2003, Brief Analysis of Green Ecological Port[J]. *Yangtze River*. **(5)**: 28-29

IOP Conf. Series: Earth and Environmental Science **133** (2018) 012033 doi:10.1088/1755-1315/133/1/012033

Uncertainty Evaluation of Residential Central Air-conditioning Test System

Haoxue Li

Department of Aviation Manufacture, Shanghai Civil Aviation College, No. 1, West Longhua Road, Xuhui District, Shanghai, China

Abstract. According to national standards, property tests of air-conditioning are required. However, test results could be influenced by the precision of apparatus or measure errors. Therefore, uncertainty evaluation of property tests should be conducted. In this paper, the uncertainties are calculated on the property tests of Xinfei13.6 kW residential central air-conditioning. The evaluation result shows that the property tests are credible.

1. Introduction

In recent years, residential central air-conditionings have been widely used. According to *Water chilling (heat pump) packages for household and similar application* (GB18430.2)[1], specified percentage of products must be tested under standardized condition. Moreover, *The minimum allowable values of the energy efficiency and energy efficiency grades for unitary air conditioners* (GB19576)[2] stipulates new national requirements on energy efficiency limits of air-conditioning equipment. As required by these two regulations, manufacturers should build air-conditioning property testing laboratories following the requirements of *General requirements for the competence of testing and calibration laboratories* (GB/T15481)[3], and should conduct property tests on air-conditioning units. Since the testing accuracy could influence the test result directly [4], we should also assess the uncertainties in accordance with *Evaluation and Expression of Uncertainty in Measurement* (JJF1059)[5]. However, even though all air-conditioning factories are capable of conducting property tests of air-conditioning, the uncertainty evaluation are not performed in some factories, which may cause the tests remain unreliable.

2. Measure precision analysis

2.1. Test requirements

Residential central air-conditionings properties should be tested according to *The methods of performance test for positive displacement & centrifugal water-chilling units and heat pump* (GB/T10870) [6]. All apparatuses and equipment should be within their effective checking period, with all indicators of apparatuses meeting relative specifications.

While the power of air-conditioning could be calculated, in this test the power value is acquired by measurement.

The uncertainties evaluation process was conducted on XFHLR13.6 Unit (Xinfei 13.6kW air cooled heat pump residential central air-conditioning) as shown in Fig.1. Fig.2 shows the property test system.

Figure 1. XFHLR13.6L Unit

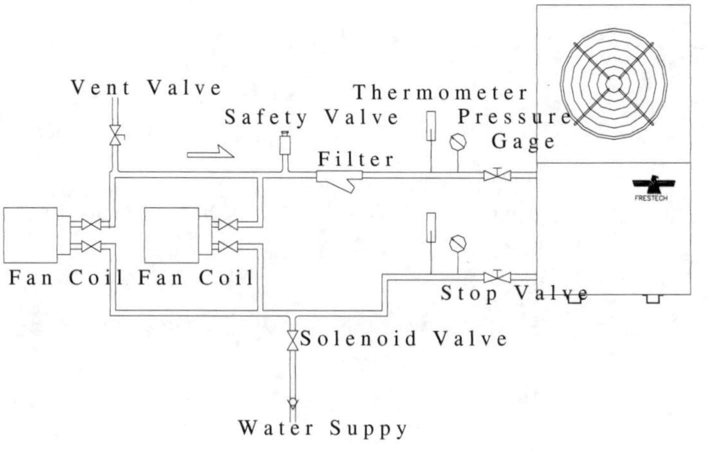

Figure 2. Properties Test System

2.2. Test errors and precision

As the level of precision relates to error, error could be used as a measurement of the level of precision. The smaller the error, the higher the precision level is.

While error is a determined value, the truth value is an ideal concept. The truth value refers to the measure value when certain value can be determined and all limits are eliminated. When processing data, the truth value should be replaced by conventional truth value.

Uncertainties refer to the range of measure error obtained in the evaluation of the magnitude range of the measured truth value. To be specific, uncertainties show the comprehensive distribute range of the random error and uncertain system error. It could be interpreted as the value of error under certain confidence probability. As they are important indicators of measurement quality, uncertainties should be given when physical value measurement data is provided, so that users could evaluate the reliability. Uncertainty should be evaluated at appropriate level of precision. Higher precision level may lead to suspicion of the test result validity. Otherwise, it may lead to wrong conclusion and may potentially affect the quality assurance of products.

2.3. Measure precision analysis

Regarding the actual measures, high level of exactitude does not necessarily mean high degree of veracity. Likewise, high degree of veracity does not necessarily mean high level exactitude as well. However, high level of accuracy means high level of exactitude and high degree of veracity.

The quantitative characteristics of the accuracy can be expressed by uncertainty in measurement.

The sources of uncertainty measurement include imperfection of measured amount, non-ideal measure method and sampling, artificial deviation in apparatus indicator reading, uncertainty of measure standard and the localization of measure instrument.

3. Evaluation of uncertainty in measurement[7]

Guide to the Expression of Uncertainty in Measurement [8] was issued in 1993 by ISO and other six international organizations. *Evaluation and Expression of Uncertainty in* Measurement (JJF1059-1999) and *General requirements for the competence of testing and calibration laboratories* (GB/T15481-2000) were issued in China 6 years later, which request that testing laboratories should have the capabilities of evaluating uncertainties in measurement.

Standard uncertainty refers to the uncertainty in measurement shown as standard deviation. Related uncertainty refers to the ratio of uncertainty in measurement to true value.

In accordance with JJF1059-1999, the evaluation process of uncertainty in measurement is shown in Fig.3.

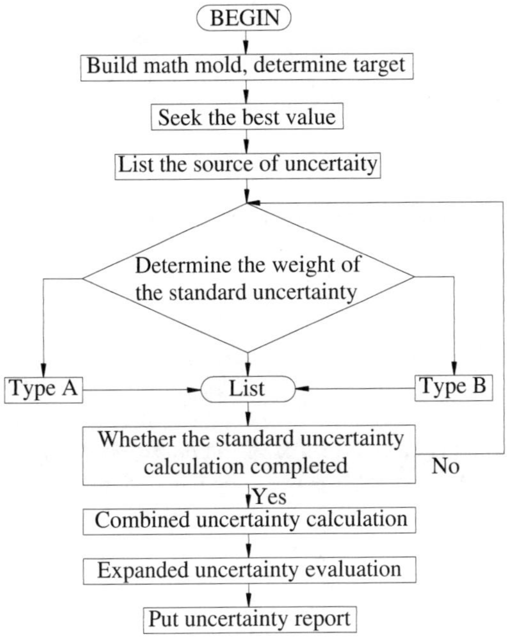

Figure 3. The evaluation process of the uncertainty in measurement

3.1. Estimation of uncertainty in measurement

Uncertainty is caused by error. Due to the complexity of error, it is difficult to calculate the uncertainty accurately. The estimation method is often used as follows.

$$U = \sqrt{U_A^2 + U_B^2} \tag{1}$$

U_A -Type A standard uncertainty, acquired by statistical method.

U_B -Type B standard uncertainty, acquired by non-statistical method.

Type A component estimate:

When an independent variable x_i is the result of the arithmetic mean $\overline{x_i}$ in n times repeat measurements, the standard uncertainty is the standard deviation of $\overline{x_i}$.

$$U_A = \sqrt{\frac{\sum_i^n \left(x_i - \bar{x}\right)^2}{n(n-1)}} \tag{2}$$

Type B component estimate:

Type B standard uncertainty can be evaluated using different information sources under relative conversion relationship. In this test, x_i is obtained from the instruction book or operation manuals of the instruments. The expanded uncertainty U and its coverage factor d are known, then:

$$U_B = \frac{U}{d} \tag{3}$$

If the measured estimate value obeys uniform distribution, then:

$$d = \sqrt{3} \tag{4}$$

If the measured estimate value obeys inverse sine distribution, then:

$$d = \sqrt{2} \tag{5}$$

3.2. Uncertainty in measurement evaluation

The ranges and accurate degree of the main measurement instrument are shown in Table 1.

Table 1. Range and accuracy degree of the instrument

No.	Instrument	Type	Serial number	Range	Accuracy degree
1	Poly temperature tester	TC-2	180231	-50~250 ℃	0.05 ℃
2	Turbine flowmeter	LWGY-25	04-084877 04-089212	1~10 m^3/h	±0.5%
3	Electricity tester	AN7931A	047903072 047903073	Voltage: AC10.0V~600.0V Current: 0.030~40.00A Power: 0.001~72kW	0.25% × (Indication +Range)

3.2.1. The power measurement uncertainty

Type A uncertainty:

Power varies with the environment temperature in varying duty tests. The independent variable cannot be evaluated with statistic analytical method. Thus the type A uncertainty is non-existent, that is:

$$U_A = 0 \tag{6}$$

Type B uncertainty:

According to Table 1, then:

$$\Delta_I = 0.25\% \times \text{Indication} + 0.25\% \times \text{Range} = 0.25\% \times 6.28 + 0.25\% \times 72 = 0.196\text{kW} \tag{7}$$

(The maximum power value is 6.28 kW)

If power measurement obey uniform distributing, then:

$$U_B \leq \frac{\Delta_{仪}}{\sqrt{3}} = \frac{0.196}{\sqrt{3}} = 0.113 \text{ kW} \tag{8}$$

The combined uncertainty is:

$$U = \sqrt{U_A^2 + U_B^2} = U_B = 0.113\text{kW} \tag{9}$$

3.2.2. The temperature measurement uncertainty

Type A uncertainty:

Same as the power measurement:

$$U_A = 0 \tag{10}$$

According to Table 1:

$$\Delta_I = 0.05 \text{ ℃} \tag{11}$$

Type B uncertainty:

If temperature measurement obeys uniform distribution, then:

$$U_B \leq \frac{\Delta_I}{\sqrt{3}} = \frac{0.05}{\sqrt{3}} = 0.029\text{℃} \tag{12}$$

The combined uncertainty is:

$$U = U_B = 0.029\text{℃} \tag{13}$$

3.2.3. The flux measurement uncertainty

Same as the power measurement:

$$U_A = 0 \tag{14}$$

Type B uncertainty:

According to Table 1:

$$\Delta_I = 10 \times 0.5\% = 0.05\text{m}^3/\text{h} \tag{15}$$

$$U_B \leq \frac{\Delta_I}{\sqrt{3}} = \frac{0.05}{\sqrt{3}} = 0.029\text{m}^3/\text{h} \tag{16}$$

The combined uncertainty is:

$$U = U_B = 0.029\text{m}^3/\text{h} \tag{17}$$

3.3. Related uncertainty in measurement valuation

The standard condition test result is shown in Table 2.

Table 2. Test parameters

No.	Content	Refrigeration	Heating
1	Water flux	0.626	0.688
2	Δt, ℃	5	5
3		13.10	15.37
4	Input power, kW	4.87	4.92
5	COP	2.69	3.12
6	Freon high pressure	1.82	1.80
7	Freon low pressure	0.41	0.32
8	Compress ratio	4.44	5.63
9	Power factor	83.6%	86.3%

When the above values are taken as true values, the related uncertainties in measure of each parameter are as follows:

$$\text{Power: } U = 0.113/4.87 = 2.32\% \tag{18}$$

$$\text{Temperature: } U = 0.029/7 = 0.41\% \tag{19}$$

$$\text{Flux: } U = 0.029/2.25 = 1.29\% \tag{20}$$

4. Conclusion

Based on the calculation of uncertainties in measurement, a conclusion could be reached: for the residential central air-conditioning property tests, all the related uncertainties in measurement of the measurement parameters are less than 2.5%. So the test results are true and reliable.

5. Acknowledgments

Regarding the support to this article, many thanks to Xinfei Central Air-conditioning Company.

6. References

[1] GB18430.2-2001, Household and similar water chilling (heat pump) packages, China Standard Press, 2001

[2] GB19576-2004, The minimum allowable value of the energy efficiency and energy efficiency grades for unitary conditioners, China Standard Press, 2004

[3] GB/T15481-2000, General requirements for the competence of testing and calibration laboratories, China Standard Press, 1999

[4] Wang Ziting, Thermal Energy and Power Test Technology, Xian Jiaotong University Press, 1998, pp 1-15, 82-173

[5] JJF1059-1999, Evaluation and Expression of Uncertainty in Measurement, China Measure Press, 1999

[6] GB/T10870-2001, The methods of performance test for positive displacement water-chilling units and heat pump, China Standard Press, 2001

[7] Fei Yetai, Error Theory and Date Processing, China Machine Press, 2005, pp 1-93

[8] ISO, BIPM, IEC, IFCC, IUPAC,OIML, Guide to the Expression of Uncertainty in Measurement